D1067853

Politics Across the Hudson

Rivergate Regionals

Rivergate Regionals is a collection of books published by Rutgers University Press focusing on New Jersey and the surrounding area. Since its founding in 1936, Rutgers University Press has been devoted to serving the people of New Jersey and this collection solidifies that tradition. The books in the Rivergate Regionals Collection explore history, politics, nature and the environment, recreation, sports, health and medicine, and the arts. By incorporating the collection within the larger Rutgers University Press editorial program, the Rivergate Regionals Collection enhances our commitment to publishing the best books about our great state and the surrounding region.

Politics Across the Hudson

The Tappan Zee Megaproject

PHILIP MARK PLOTCH

Rutgers University Press

New Brunswick, New Jersey, and London

Library of Congress Cataloging-in-Publication Data
Plotch, Philip Mark, 1961–
Politics across the Hudson : the Tappan Zee megaproject / Philip Mark Plotch.
pages cm. — (Rivergate regionals)
Includes bibliographical references and index.
ISBN 978–0–8135–7249–9 (hardback) — ISBN 978–0–8135–7251–2 (e-book (epub)) —
ISBN 978–0–8135–7252–9 (e-book (Web PDF))
1. Tappan Zee Bridge (N.Y.) 2. Transportation—Political aspects—New York Metropolitan Area. 3. Local transit—Political aspects—New York Metropolitan Area. 4. Transportation and state—New York Metropolitan Area. I. Title.
TA1025.N49P56 2015
388.1'3209747277—dc23 2014035983

Visit our website: http://rutgerspress.rutgers.edu

Manufactured in the United States of America

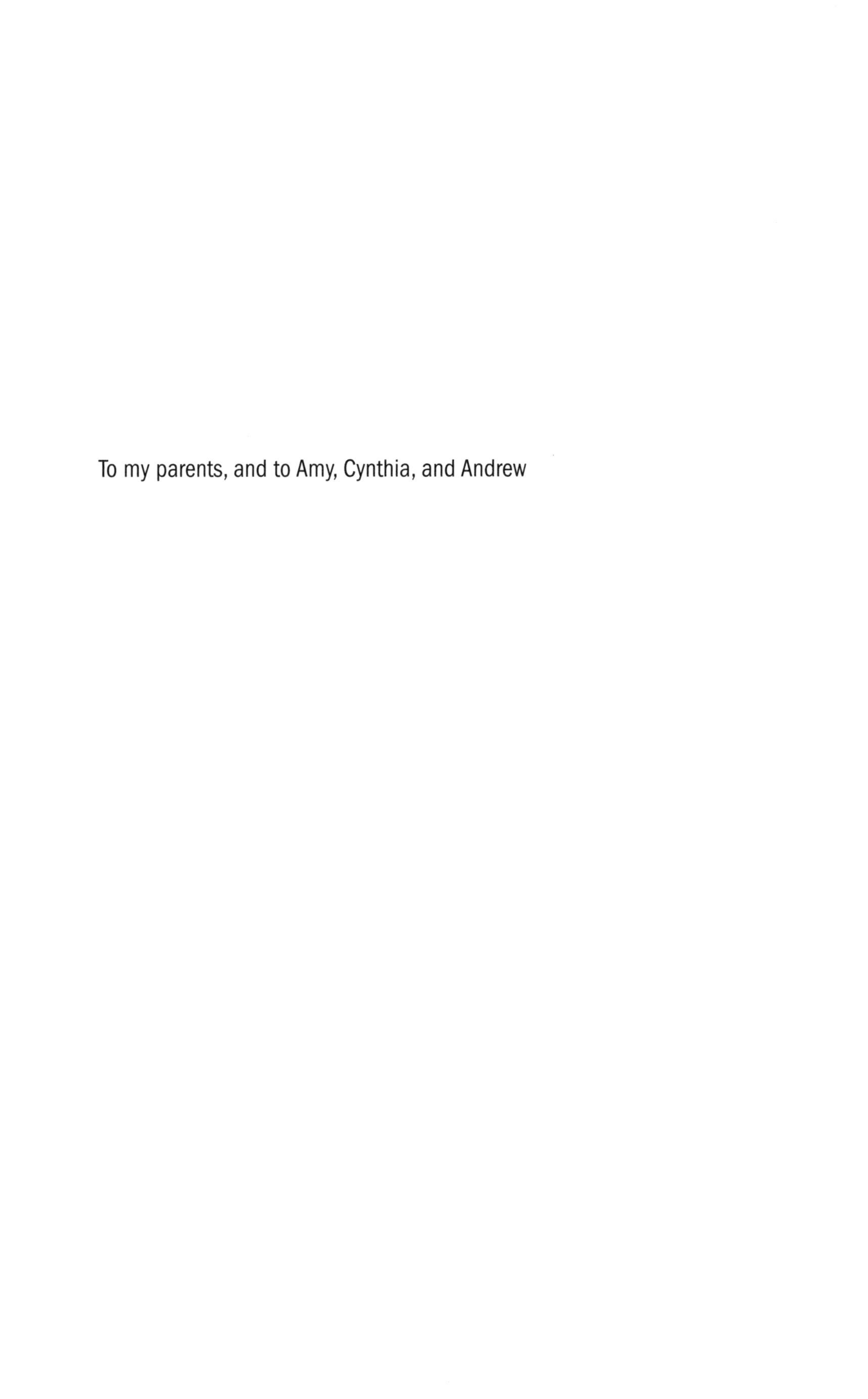

To my parents, and to Amy, Cynthia, and Andrew

Contents

Illustrations

Tables

Acknowledgments

First, I must thank Hunter College's Bill Milczarski and the New School's Lisa Servon who provided me with years of guidance, encouragement, and support. Academics who helped me place this saga into a broader context include the City College of New York's Buz Paaswell, the New School's David Howell, Rick McGahey, Jeff Smith, Rachel Meltzer, and Erica Kohl-Arenas, as well as Hunter College's Arielle Goldberg, Owen Gutfreund, and Joseph Viteritti. I also must pay tribute to two brilliant authors, Robert Caro and Anthony Downs, who have provided me with invaluable insight into politics, economics, metropolitan governments, sprawl and transportation.

I am grateful to many people who transformed my words into a published book. Michael Roney helped me turn a fascinating story into a compelling manuscript, Sarah Clarehart transformed a thousand of my words into a single figure, and Amy Sutnick Plotch helped me every step of the way. At Rutgers University Press, Marlie Wasserman's enthusiasm ensured that the manuscript obtained all the resources it needed, Marilyn Campbell choreographed the production, and Willa Speiser smoothed the rough edges.

My research was like putting together a giant puzzle, and I appreciate more than a hundred people who gave me their puzzle pieces, especially Janine Bauer, Peter Derrick, Naomi Klein, Janet Mainiero, Maureen Morgan, Rich Peters, and Jeff Zupan.

I had the great fortune to interview U.S. Department of Transportation Secretary Raymond LaHood and New York governors George Pataki, David Paterson, and Eliot Spitzer. They helped me understand the political dynamics in the White House and Albany's executive mansion. Agency heads and governors' aides offered me invaluable insight into the pressures faced by the men and women responsible for improving New York's extraordinarily complex transportation system. I would like to thank a dozen of them—John Cahill,

Virgil Conway, Tim Gilchrist, Maryanne Gridley, Stanley Kramer, Carrie Laney, Charles Lattuca, Steve Morgan, Richard Ravitch, Elliot Sander, John Shafer, and Lou Tomson.

I sincerely appreciate all the devoted civic advocates who told me their stories, including Greg Clary, Orrin Getz, George Haikalas, Marlene Kleiner, Elyse Knight, Charles Komanoff, Mark Kulewicz, Goodie Lelash, Nicole Gelinas, Ross Pepe, Michael Replogle, Alex Saunders, Al Samuels, Sy Schulman, Kate Slevin, Jim Tripp, Veronica Vanterpool, and Robert Weinberg.

Officials at the Thruway Authority, Metropolitan Transportation Authority (MTA), and the New York State Department of Transportation (NYS DOT) have devoted their careers to improving the state's transportation system. I appreciate their efforts and the conversations I had with Michael Anderson, Bill Aston, Jim Barry, Al Bauman, Dan Coots, Larry DeCosmo, Gary Dellaverson, Bob Dennison, Leonard DePrima, Peter Derrick, Duane Dodds, Larry Dwyer, Dan Evans, Keith Giles, Tony Gregory, Mark Herbst, Steve Herrmann, Marty Huss, Tony Japha, Angel Medina, Peter Melewski, Richard Newhouse, Ted Orosz, Joe Pasanello, Henry Peyrebrune, Sarah Rios, Lou Rossi, Jean Shanahan, Brian Sterman, Wayne Ugolik, Elisa Van Der Linde, Chris Waite, Alan Warde, Bill Wheeler, and Robert Zerrillo.

Other government officials and consultants who generously shared their perspectives included Jerry Bogacz, Ed Buroughs, Chris Calvert, Patty Chemka, Robert Conway, Todd Discala, Peter Feroe, Drew Fixell, Patrick Gerdin, Dan Graves, Dan Greenbaum, Tom Harknett, Jim Hartwick, Terry Hekker, Sandy Hornick, Irwin Kessman, Jay Krantz, Thomas Parody, James Parsons, George Paschalis, Marty Robins, David Rubin, Ernie Salerno, Tom Schulze, Neil Trenk, Scott Vanderhoef, and James Yarmus.

I would also like to offer a special acknowledgment to many key officials who shared information about the I-287 planning process yet wish to remain anonymous. To protect their confidentiality, I have not included the location and dates of my interviews with them. Finally, I owe a very special thanks to my anonymous friend in the governor's office and my two favorite transportation consultants.

Guides to This Book

Because the improvements planned along the I-287 corridor were poised to reshape the entire New York metropolitan area, hundreds of elected officials, bureaucrats, business leaders, grassroots activists, and everyday citizens participated in the planning process. The following figures and tables provide a quick reference for readers to recall the major players, institutions, and events.

- Table G.1 lists key organizations.
- Table G.2 lists key players in the planning process.
- Figure G.1 shows the tristate New York metropolitan area. (The I-287 corridor in New York runs across Rockland and Westchester Counties.)
- Figure G.2 shows the east-west I-287 corridor as well as the existing rail lines that connect the corridor with northern New Jersey (west of the Hudson River) and Manhattan (east of the Hudson River).
- Figure G.3 provides a timeline of key events along with the tenure of six governors and three Westchester County executives.

Table G.1
Key Organizations

Organization	Role
Federated Conservationists of Westchester County	Coalition of Westchester County environmental groups.
New York State Department of Transportation (NYS DOT)	State agency responsible for most of New York's highways.

(*continued*)

Table G.1 (*continued*)

Metropolitan Transportation Authority (MTA)	Largest transportation provider in the U.S., whose subsidiaries include Long Island Rail Road, Metro-North Railroad, New York City Transit, and MTA Bridges and Tunnels.
Metro-North Railroad	Subsidiary of the Metropolitan Transportation Authority that provides passenger railroad services to New York City and its northern suburbs.
New York State (NYS) Thruway Authority	Owner and operator of the New York State Thruway, including the Tappan Zee Bridge.
New York Metropolitan Transportation Council	Metropolitan planning organization that programs all federal transportation funds for projects in New York City, Long Island, and Rockland, Westchester, and Putnam Counties.
Port Authority of New York and New Jersey ("Port Authority")	Bi-state transportation agency responsible for Hudson River crossings, airports, and transit services. It was originally known as the Port of New York Authority.
Regional Plan Association	Independent, not-for-profit regional planning organization that recommends improvements for the New York metropolitan area.
Tri-State Transportation Campaign ("Tri-State")	Advocacy group established in the 1990s to reduce car dependency in New York, New Jersey, and Connecticut.
U.S. Department of Transportation	Provides federal transportation funding to states and localities. Its subsidiaries include the Federal Transit Administration and the Federal Highway Administration.
Westchester County Association	Westchester County business organization.

Table G.2
Key Players

Name	Position
Michael Anderson	NYS DOT engineer and manager of the I-287 / Tappan Zee project
Janine Bauer	Tri-State Transportation Campaign executive director
Andrew Cuomo	Governor of New York
George Case	Federated Conservationists of Westchester County president
Robert Conway	AKRF (environmental consulting firm) engineer
Virgil Conway	MTA chairman, I-287 Task Force chairman, and NYS Thruway Authority board member
Robert Dennison	NYS DOT regional director and chief engineer

Table G.2 (*continued*)

Thomas Dewey	Governor of New York
Drew Fixell	Tarrytown mayor
Tim Gilchrist	NYS DOT executive and senior advisor to Governors Spitzer and Paterson
Janet Mainiero	Metro-North planner
Maureen Morgan	Federated Conservationists of Westchester County president and Westchester County Chamber of Commerce official
Andrew O'Rourke	Westchester County Executive
George Pataki	State legislator and governor of New York
David Paterson	Lieutenant governor and governor of New York
Ross Pepe	New York Construction Industry Council of Westchester and Hudson Valley president
Howard Permut	Metro-North's planning director, vice president, senior vice president, and president
Rich Peters	NYS DOT engineer
John Platt	NYS Thruway Authority executive director
Richard Ravitch	MTA chairman and Governor Paterson's lieutenant governor
Al Samuels	Rockland Business Association president
Sy Schulman	Westchester County planning commissioner and president of the Westchester County Association
John Shafer	NYS Thruway Authority executive director
Kate Slevin	Tri-State Transportation Campaign executive director
Lou Tomson	NYS Thruway Authority chairman and Governor Pataki's deputy secretary
Jim Tripp	Environmental Defense Fund attorney and Tri-State Transportation Campaign founder
Scott Vanderhoef	Rockland County Executive
Chris Waite	NYS Thruway Authority chief engineer
Robert Weinberg	Westchester County commercial real estate developer
Franklin E. White	NYS DOT Commissioner
Jeff Zupan	Regional Plan Association senior fellow

FIGURE G.1 *Above*: The tristate New York metropolitan area is home to approximately 22 million people. The I-287 corridor runs across Rockland and Westchester Counties in New York.

Source: NYCRuss, "Regional Plan Association 31 County Area," http://en.wikipedia.org/wiki/File:Regional_Plan_Association_31_County_Area.png. Creative commons license.

FIGURE G.2 *Facing page*: The east-west I-287 highway includes the Tappan Zee Bridge across the Hudson River. Metro-North provides train service from New York's northern suburbs on the Port Jervis and Pascack Valley Lines to northern New Jersey (west of the river) and on the Hudson, Harlem, and New Haven Lines to Manhattan (east of the river).

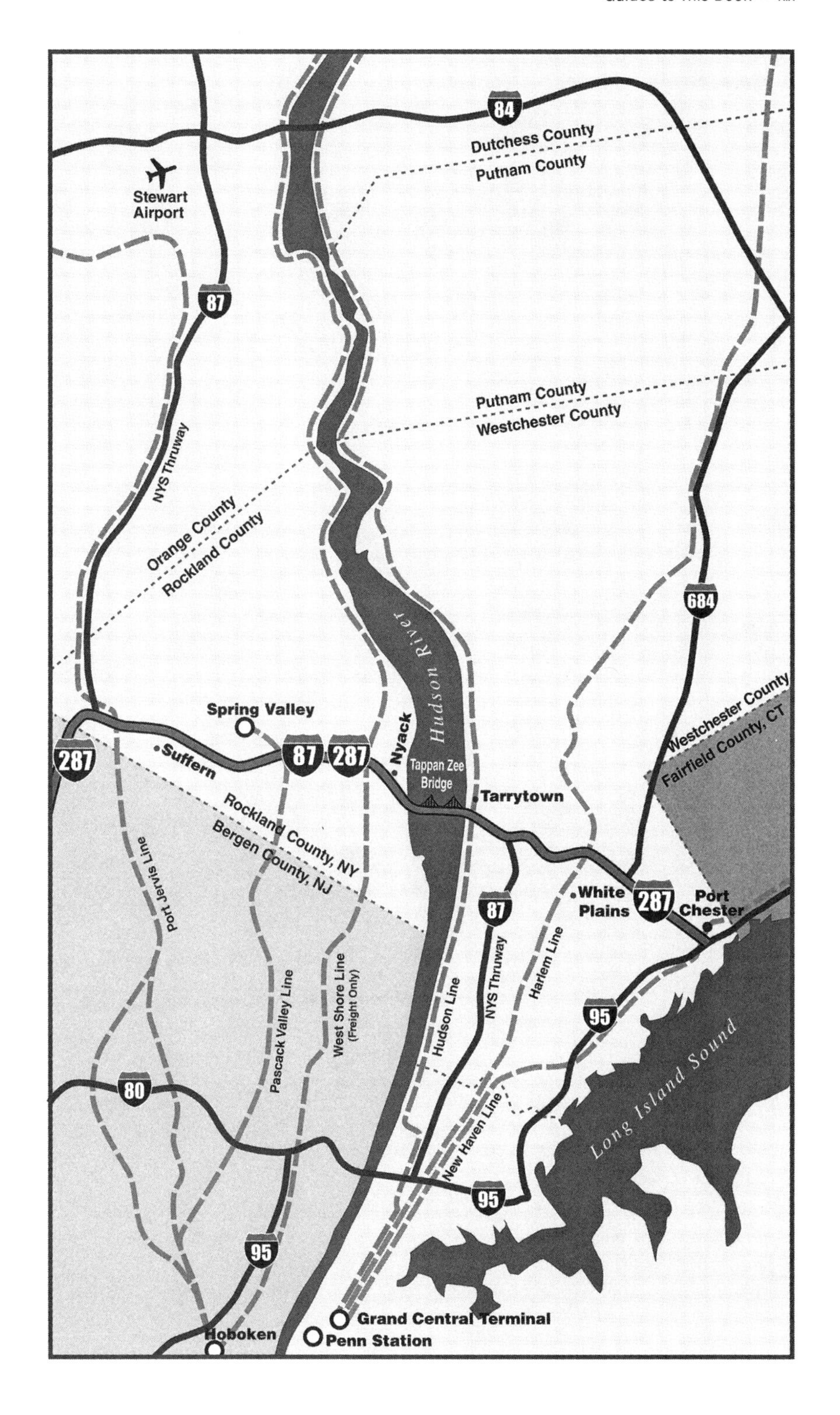
84
Dutchess County
Putnam County
Stewart
Airport
87
Putnam County
Westchester County
NYS Thruway
Orange County
Rockland County
Hudson River
684
Westchester County
Fairfield County, CT
Spring Valley
287
Suffern
87
287
Nyack
Tappan Zee
Bridge
Tarrytown
Rockland County, NY
Bergen County, NJ
White
Plains
287
Port
Chester
87
Port Jervis Line
Pascack Valley Line
West Shore Line
(Freight Only)
Hudson Line
NYS Thruway
Harlem Line
95
Long Island Sound
80
New Haven Line
95
95
Grand Central Terminal
Hoboken
Penn Station

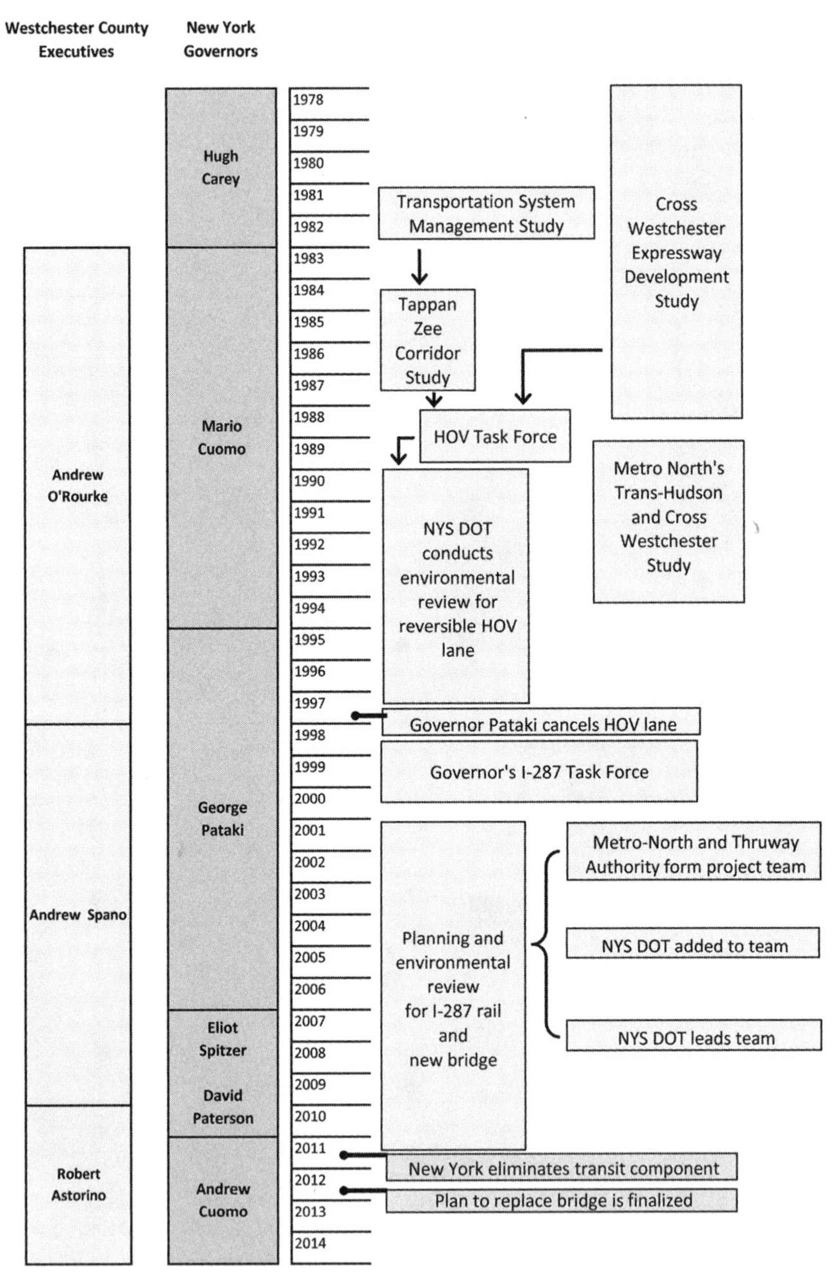

FIGURE G.3 This timeline shows the key events of the I-287/Tappan Zee Bridge megaproject planning process spanning the tenure of six New York governors and three Westchester County executives.

Politics Across the Hudson

Introduction

Large-scale infrastructure projects—megaprojects—can fuel economic growth. New York grew from a colonial outpost into a world capital because of strategic investments in canals, railroads, subways, tunnels, highways, bridges, and airports. More recently, America's failure to invest in its transportation infrastructure has led to congestion and capacity constraints that threaten to impede its global competitiveness.[1] Today, one quarter of America's bridges are functionally obsolete,[2] states have stymied efforts to create a high-speed rail network, and the government has not allocated sufficient resources to adequately maintain, let alone significantly improve, the transportation network.[3]

Many government attempts to expand capacity and address regional congestion have taken decades and consumed vast resources with little to show for them. Maryland had to wait twenty-seven years until the federal government approved its environmental review documents for a new highway, the InterCounty Connector.[4] Virginia residents have waited even longer for government approvals of the Charlottesville Bypass, a proposed six-mile road designed to circumvent thirteen traffic signals.[5] New Yorkers have been the most patient, though. They have been waiting nearly one hundred years for a subway line to be built under Second Avenue in Manhattan.

States do sometimes overcome the daunting obstacles that thwart the successful planning and construction of megaprojects. The New York State Thruway Authority is now replacing its three-mile-long Tappan Zee Bridge across the Hudson River. Located about ten miles north of New York City, two new spans will create one of the world's longest and widest bridges.

At the opening ceremonies for the new bridge, elected officials will undoubtedly praise the construction workers, transportation agencies, and each other for completing the multibillion-dollar project. The spectators will be witnesses not only to the inauguration of a public sector triumph but also to the culmination of a failure-ridden process.

New York wasted more than three decades trying to finalize plans for this section of the I-287 highway corridor. If the metropolitan area's leaders had cooperated on implementing improvements during that time, they could have improved the region's transportation infrastructure much more dramatically. Instead, the region abandoned viable options, squandered hundreds of millions of dollars, forfeited more than three billion dollars in federal funds, and missed out on important opportunities.

If the state had tried simply to replace a bridge, it could have accomplished that decades ago. However, government officials had much more ambitious goals for the I-287 corridor. They tried to solve a congestion problem that could not be solved, while inadvertently triggering hot-button issues relating to economic development, sprawl, housing, race, urban decline, local autonomy, and long-term sustainability. In 2009, a leading transportation scholar, Robert Paaswell, described the ongoing debate about the corridor as the "quintessential 21st-century planning dilemma."[6]

For three decades, transportation officials were unable to develop a consensus on a realistic project. The reasons were numerous. The State of New York lacked sufficient funds to meet the expectations of an affluent and influential community, and the federal government was unable to fill the funding gap since it has not raised the gasoline tax since 1993. Transportation officials also faced a complex set of environmental and planning regulations that were not in place in 1964, when New York built its last major new crossing, the Verrazano-Narrows Bridge.

Furthermore, various players from the private, public, and civic sectors raised false expectations about the feasibility of certain transportation options and the state's ability to solve recurring traffic congestion. Planning was repeatedly delayed when government officials went behind each other's backs or simply were unable to work together. These officials had different interests, competed for scarce resources, and had conflicting ideas on how to address the I-287 congestion problem.

Three governors preferred to keep studying the transportation problem rather than make a decision that would have disappointed important constituencies. Given the public's high expectations, it seemed impossible for them to identify a solution that was both practical and popular. They had little to gain from lowering expectations; instead, they were trapped by them.

Institutional barriers have long been blamed for the inability of metropolitan areas to address transportation problems. In 1965, New York City mayoral candidate John Lindsay bemoaned the lack of a coordinated citywide transportation policy. He said, "We can no longer afford the ineffectiveness, the quarrels, and the waste which characterize the hydra-headed and fragmented monstrosity of the city's transportation structure today."[7]

Few American metropolitan areas have governmental institutions that can effectively address region-wide congestion. A government with region-wide transportation and land use powers has a decided advantage in battling congestion problems. It can determine the location of highway and transit routes and identify the most appropriate location and density for residential and commercial developments.

In the nineteenth century, many cities had these powers because they expanded their geographical limits. Boston, New York, Philadelphia, Chicago, Pittsburgh, St. Louis, Cleveland, and Denver became ten to fifty times larger.[8] If America's older cities had continued to expand at the same rate, Boston would now extend to Route 128, Chicago would stretch half the distance to Milwaukee, and New York City would reach to I-287 in Westchester.[9] However, most cities stopped expanding in the twentieth century because wealthy suburban communities had little incentive to join forces with the major municipalities at their region's core. So, instead of very large cities or regional governments with broad powers, most metropolitan areas today are fragmented into scores of municipalities and separate government jurisdictions, each with their own unique responsibilities for transportation and land-use decisions.

New York may be the most difficult place in the nation to undertake regional planning efforts. After all, it is one of the most politically fragmented metropolitan areas in the Western world, with a high degree of both functional and geographical fragmentation. A former senior state transportation official told me that with all the transportation agencies and institutional problems, "you couldn't design a worse scenario."[10]

The New York region's transportation system is dominated by quasi-governmental institutions that include the New York State Thruway Authority, Metropolitan Transportation Authority, Port Authority of New York and New Jersey, New Jersey Turnpike Authority, and the New Jersey Transit Corporation. Harvard University's Gerald Frug finds that the widespread use of authorities and government-owned corporations, combined with numerous municipal governments, makes regional coordination and planning virtually impossible.[11]

The metropolitan area's three states—New York, New Jersey, and Connecticut—tend to compete more than cooperate. In 2010, when New York governor David Paterson informally reached out to New Jersey governor Chris Christie to see if he would be interested in working with New York to replace the aging Tappan Zee Bridge, Christie responded, "Stop screwing with us."[12] It was the sign of a region that has been unable to establish an ongoing productive working relationship to address long-term transportation capacity and congestion issues.

Cooperation in the New York region is possible and does happen. New York and New Jersey established the Port Authority in 1921 to build infrastructure that neither state could create on its own. As a result, the New York metropolitan area now has a thriving seaport, three of the nation's busiest airports, and six highway crossings between New York City and New Jersey. The metropolitan area's transportation agencies also worked together to implement the E-ZPass system for convenient travel on all the region's toll roads. It might seem obvious to drivers that each of the tolling agencies would use the same electronic tag, but it required the agencies and states to work together, give up some control, and transcend their parochial concerns.

In the New York metropolitan area, however, no government institution looks at the region's overall transportation problems and evaluates projects that can address them. Instead of asking what the region needs to remain globally competitive, its powerful transportation agencies pursue strategies to advance their own missions.[13]

Politics Across the Hudson considers how the players who help influence megaprojects often put their own interests ahead of those of the public. Individuals in the public, private, and civic sectors have a vested interest in raising false expectations to further their own goals. When self-interested behavior is exposed, it shows people behaving in predictable ways. Governors try to get reelected, consultants seek more contracts, developers lobby for new public infrastructure, and bureaucrats try to obtain more resources for their agencies, while planners and engineers figure out how to get their projects built.

Some rather sinister-sounding terminology goes along with this behavior. The dark side of America's suburban development has been sprawl and widespread traffic congestion. Planners refer to these problems as *wicked* because they resist permanent solutions. Likewise, the dark side of the quasi-public institutions that build and maintain transportation infrastructure is the way they distort the public agenda to serve their own needs, and the dark side of forecasting is the way project sponsors often overestimate benefits and underestimate costs in order to increase the likelihood of getting their projects built.[14]

Politics Across the Hudson describes various types of politics, from promises on the campaign trail to bureaucrats playing political games. This behind-the-scenes look at three decades of planning and politics may confirm your worst fears about governmental dysfunction, but it also reveals essential lessons for those interested in tackling today's complex public policy problems. This book examines the factors that delay the planning of America's infrastructure projects, how obstacles can be overcome, and the opportunities that can be lost along the way.

1

The I-287 Corridor

From Conception to Congestion

New York City has one of the world's greatest natural harbors. The water is deep, the currents are mild, and it sits at the mouth of the 315-mile-long Hudson River. In the nation's early years, three other well-sited cities—Boston, Philadelphia, and Charleston—had busier ports.[1] The fact that New York's population and economy far surpass every other American city today is a result of one of the most successful economic development initiatives ever—the $7 million 363-mile-long Erie Canal, which connected the Great Lakes and the Hudson River.

In the early nineteenth century, New York governor DeWitt Clinton predicted that after the Erie Canal opened, New York City would "become the granary of the world, the emporium of commerce, the seat of manufactures, the focus of great moneyed operations." A few years after the canal opened in 1825, the governor's predictions came true. New York's port was handling more goods than the next three cities combined.[2] Its rise as a global trading center spawned the city's prominence as the nation's manufacturing, financial, corporate, and media center.

The canal also transformed the rest of the state. With the exception of Binghamton and Elmira, every major city in New York falls along the trade route established by the Erie Canal, from New York City to Albany, through Schenectady, Utica, and Syracuse, to Rochester and Buffalo.[3]

After the Erie Canal opened, New York's transportation investments continued to feed its growth. Beginning in the 1840s, railroad lines to New York City were built along Westchester County's Hudson River, Saw Mill River, Bronx River, and Long Island Sound. The railroad attracted industry and residents to new communities around Westchester's rail stations. By the late nineteenth century, Westchester had become the first large-scale suburban area in the world, with upper-middle-class communities in Scarsdale, Mount Vernon, New Rochelle, Bronxville, and Rye.[4] Its residents relied on the railroads as well as extensive trolley systems in many of its cities.

Westchester had another growth spurt after three parkways opened in the late 1920s and early 1930s. These meandering scenic highways were designed to connect New York City with its countryside and state parks, and like the railroad and the rivers, were oriented in a north-south direction. The parkways also opened up tracts for residential development and led to a population increase of more than 50 percent during the 1920s.

Rockland County, across the river from Westchester, did not experience the same type of growth. It had no direct rail connections to New York City and it did not get a parkway until after World War II. In 1940, Westchester had over 573,000 residents while Rockland had less than 75,000. Rockland was still a semirural countryside with compact villages and hamlets separated by forests, meadows, and farms.[5]

In the late 1940s, New York's governor Thomas E. Dewey predicted that another transportation improvement would transform the state's economy as the canals and railroads had in previous generations. He wanted to build "the greatest highway in the world."[6]

Years before the development of the nation's interstate highway system, Dewey envisioned a five-hundred-mile long superhighway that would connect nearly all of the state's population centers. Running relatively parallel to the Erie Canal and the southern portion of the Hudson River, the highway would have no stoplights, no crossings, and no grades exceeding 3 percent. At the Thruway's groundbreaking in 1946, Dewey said that it would "create a growing society, happier travel, greater safety and access for every New Yorker to the other parts of the state."[7]

Building a new highway was a much easier proposition in Dewey's time than it is today. The New York metropolitan area still had wide-open spaces with relatively inexpensive land. The federal government had only a limited oversight role, the public was not environmentally sensitive, and little public outreach was required. However, the highway's construction was delayed because of a lack of funding. In early 1950, Dewey broke the financial logjam when he decided that the Thruway would be a toll road, financed and operated by an independent authority.

That spring Dewey signed legislation to set up the New York State Thruway Authority, a quasi-independent state organization that would finance construction by issuing bonds to be paid back with toll revenues.[8] Dewey liked public authorities because they could finance costly projects, allowing the state to avoid raising taxes or waiting for voters to pass a referendum to increase the state's debt ceiling. For elected officials, public authorities provide an effective way to gain short-term political benefits with minimal short-term costs.

Advocates for authorities claimed they could build projects faster and more efficiently. With their own operating procedures and sources of revenue, they could also avoid bureaucratic civil service requirements and the vagaries of the state's annual budget process.[9] By the 1950s, New York had set up authorities to build parkways, bridges, tunnels, parking structures, dormitories, sewers, and other infrastructure. In fact, the New York State Thruway Authority was New York's forty-fifth such entity.[10] The state's first authority, the Port of New York Authority (popularly called the Port Authority), was established with New Jersey in 1921.

Once the Thruway Authority was established, the project proceeded expeditiously, but there was still one major outstanding issue: if and how it would connect with New York City.

The Birth of the Tappan Zee Bridge

Initial plans for the New York State Thruway called for a superhighway from Buffalo to western Rockland County along the southern border of the state, where it then would connect with New Jersey highways.[11] Drivers headed to New York City would have to cross the George Washington Bridge from New Jersey.

However, in February 1950, Governor Dewey announced a change of plans. The proposed Thruway would be built from Buffalo all the way to New York City. This meant that the state would need to build a new bridge over the Hudson River.[12] (Figure 1.1 shows the main line of the Thruway.)

By early 1950, engineers had suggested sites for a new bridge where the river is less than a mile wide between New Jersey and New York. However, Dewey rejected these locations because they fell within the jurisdiction of the bistate Port Authority, which had the exclusive right to build and operate all bridges and tunnels across the Hudson River within twenty-five miles of the Statue of Liberty.[13] The state needed the bridge's toll revenues to help support the rest of the Thruway system.[14] If a new bridge were built in the Port Authority's jurisdiction, New York would have to share its toll revenues with New Jersey.

Port Authority officials told the Thruway Authority that they could not waive the jurisdiction because it was part of its covenants with bondholders,[15] which assured that no one could build a bridge that would compete with the Port Authority's crossings and threaten the authority's ability to pay off its bonds. (Ironically, years later the Thruway Authority would point to its own bond covenants as a reason it could not accommodate requests from a New York State agency that wanted to improve transportation conditions.)

FIGURE 1.1 This map shows the route of the New York State Thruway's main line from Buffalo to New York City.

Dewey decided to build a new bridge just north of the twenty-five-mile jurisdiction, even though it was the second-widest portion of the 315-mile long river—a wide inland bay known as the Tappan Zee.[16] (Figure 1.2 shows the bridge in the context of the jurisdictional line.)

The governor, who appointed half the Port Authority's commissioners, would not even allow the authority to evaluate the feasibility of building a new bridge within its jurisdiction. New York State officials did not want any reports that would make them look foolish, since they knew it was more practical to build a bridge where the river was narrower.[17]

In April 1950, Westchester and Rockland residents were surprised to learn that the area near Nyack in Rockland County and Tarrytown in Westchester County was the Thruway Authority's first choice for a new bridge, since recent public discussions had suggested narrower sections of the river. While Thruway Authority engineers were conducting test borings in May to assess soil conditions near Nyack, the governor said that if the location was not suitable, they would keep going north until the engineers found one that was.[18] However, state officials preferred to avoid sites farther north, in part because that would require purchasing more land in affluent and densely populated Westchester County.[19]

The idea of building a bridge near Nyack was not new. In 1935, the state legislature established a Rockland-Westchester Hudson River Crossing Authority, but the that body did not get very far in planning a bridge south of Nyack once members found out about the Port Authority's jurisdiction.[20] The Rockland-Westchester Bridge Authority, created later that year, considered building a span between Nyack and Tarrytown, but after conducting test borings as far as 180 feet below the water's surface, its engineers were unable to find rock that would provide adequate foundations. The authority realized that going any deeper would make the cost of a bridge too expensive, and in 1936 it reported that such a crossing would not be built. Instead, it stated that a location within the Port Authority's jurisdiction would be more feasible.[21]

In the 1930s, New York governor Herbert H. Lehman did not try to overcome the formidable financial and logistical obstacles involved in building a

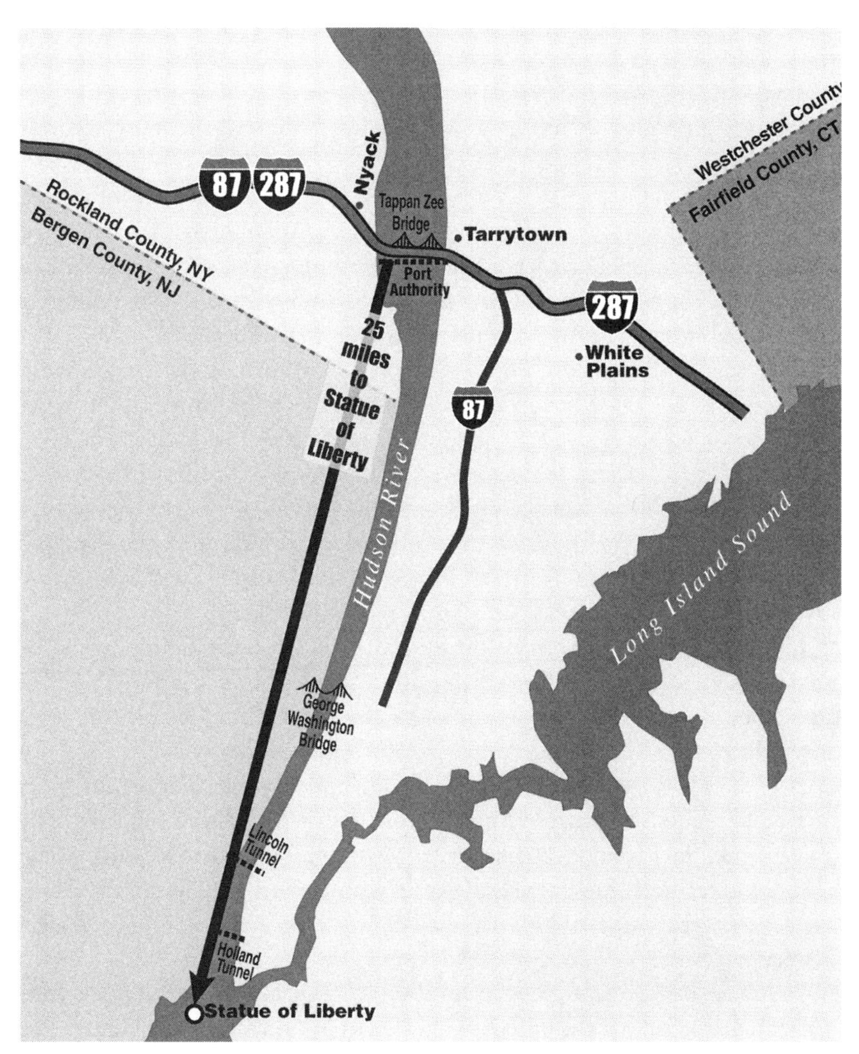

FIGURE 1.2 The dotted line below the Tappan Zee Bridge shows the northern limit of the Port Authority's twenty-five-mile jurisdiction.

bridge between Nyack and Tarrytown. In 1950, however, the state had a governor who cared deeply about the new highway and bridge, so much so that the state's superintendent of public works referred to the Thruway as Dewey's "pet project."[22]

During his 1950 reelection campaign, Dewey did not divulge the exact location of the new bridge. For months, Rockland County officials tried to obtain information about the proposed Thruway route, but the governor told the Thruway Authority to put off any announcement because it would jeopardize his reelection bid.[23] However, the governor did promise that the state would hold a public meeting to obtain community input.

Three weeks after Dewey's reelection, the state held a hearing on Manhattan's Upper East Side. Although the Thruway Authority chairman, Bertram Tallamy, still did not disclose the bridge's exact location, the *New York Times* reported that it apparently would span the river at the Tappan Zee. Referring to the Port Authority's jurisdictional limits, Tallamy said the site was "the most desirable within the limitations of law."[24]

Bridge opponents were frustrated by their inability to obtain information about its location, cost, and design. They were also annoyed that a meeting designed to solicit community input was held in New York City. Opponents accused the Thruway Authority of distorting facts when it said building the approaches for the bridge would cause a minimum of dislocations.[25]

Because the Thruway Authority officials closely guarded the details about the right-of-way for the bridge and other portions of the Thruway, they were able to fragment the opposition and minimize the time that local officials had to react.[26] They also had an effective communications strategy, with a former reporter, Robert Monahan, handling public relations. One lesson Monahan taught the Thruway Authority chairman was never to use the word "impact" because of its negative connotations. Instead, the chairman spoke about the positive aspects of construction, such as rising real estate values.[27]

In December 1950, the state finally confirmed that the bridge would be built about eight hundred feet north of the Port Authority's jurisdiction.[28] In December, Tallamy said the selection of a route through Rockland County and across the river had been under study for about six months, but that it had not been feasible for the engineers to release "preliminary information any more than a book publisher releases his first proof before the editing and correcting processes are completed."[29] Estimated to cost $50 million, the new bridge would accommodate three lanes in each direction with a center strip for disabled cars, and an emergency walkway on each side.[30] It would be one of the longest bridges in the world.

The *New York Herald Tribune* reported, "South Nyack had it worst fears confirmed. The map of the proposed new Thruway bridge over the Hudson has burst upon the village like the clap of doom, leaving a swath of desolation in its wake."[31] Three Rockland County villages (Grandview-on-Hudson, South Nyack, and Nyack) lay in the bridge's path. About 150 homes would have to be destroyed or moved.[32] South Nyack would be especially hard hit; its village officials estimated that half its properties would be removed from the tax rolls, including the Village Hall, police headquarters, three groceries, and the mayor's residence.[33]

Local opponents organized, but they had little effect on the loss of homes and businesses, the impact on river views and property values, or the removal of real estate from local tax rolls. One local official accused the Thruway Authority of saying one thing in public and offering a different version in private. Community officials also argued that the conflict between the Thruway Authority and the Port Authority should not be permitted to destroy two villages.[34]

Not only did the bridge face community hostility, but it also engendered opposition from the well-respected and highly influential Regional Plan Association, whose president, Paul Windels, was harshly critical of the proposed location.[35] Although his civic organization had promoted the construction of numerous highways and bridges in the metropolitan area, building a bridge at the Tappan Zee did not make much sense to him. In December 1950, he sent a telegram to the head of the Thruway Authority stating that final action on the proposed bridge "should be withheld until full traffic, engineering, and financial studies are made available for public information." He wrote that "the matter was too important for hurried or secret action."[36]

A few weeks later, Windels said the bridge was of a "freak design never before attempted in a structure of this length." He argued that the new bridge was "unsightly, much more costly to build than a conventional bridge, possibly a menace to navigation, apparently very badly located with respect to a maximum potential of traffic, and would create needless uproar and dislocation in the towns affected." He said, "Only the most compelling engineering and traffic necessities and total lack of alternatives could justify what has been proposed, but no such necessities have been disclosed."[37]

Windels also was critical of the future financial implications of a badly located bridge and argued "the public has a right to know just what considerations are pushing the Thruway toward this disturbing decision."[38] His concerns were ignored. Decades later, Regional Plan Association officials would again be confused and dismayed when future governors announced their own multibillion-dollar decisions regarding the bridge.

The Tappan Zee Bridge's proximity to the Port Authority's jurisdiction meant that it had to incorporate an S-curve alignment. Dan Greenbaum, who worked for the bridge's designer in the early 1950s, said, "If the alignment had been straight between the existing approach locations on either side, a section of the bridge would have encroached on the Port Authority. The curves on either side moved the alignment northward."[39]

Engineering Issues

Since the Tappan Zee Bridge was planned years before the federal government passed a series of laws protecting the environment and historical properties, the Thruway's builders faced only one hurdle from the federal government. The Rivers and Harbors Act of 1899 required authorization from the secretary of the army for the construction of any structure in or over any navigable waters of the United States. Accordingly, three weeks after the Thruway Authority released its plans for the bridge, the Army Corps of Engineers held a four-hour public hearing. There, the district engineer made clear that the agency's sole consideration in deciding whether to issue a permit was whether the bridge would cause an undue obstruction to navigable waters.

Generally speaking, Westchester County officials supported the bridge, as did the Rockland County Planning Board, while Rockland County's Board of Supervisors and the villages most heavily impacted along the river opposed it.[40] At the hearing, opponents of the bridge argued that it would not provide sufficient clearance for boats, that ice jams would form, navigation would be dangerous under fog conditions, and the river would no longer be navigable if an enemy attacked the bridge.[41]

Although most of the opponents' arguments held little weight with the Army Corps of Engineers, the Thruway Authority did make one change based on comments at the hearing. The authority had designed a 5,600-foot-long earth- and stone-filled causeway between the western shore and the bridge's main span. After residents of the area pointed out the potential for ice to form along this causeway and impede river traffic, the authority agreed to build a viaduct supported by wooden piles instead. This decision may have helped reduce ice formation, and it inadvertently helped protect the river's natural resources, but it had significant long-term repercussions. Since the river was highly polluted, the Thruway Authority decided it did not need to treat the wooden piles before they were placed into the river. In recent years, as the Hudson River has been cleaned up, a growing population of crustaceans migrating up the river has threatened to feast on the approximately twenty thousand untreated wooden piles supporting that portion of the bridge, potentially degrading the structure's stability and safety.[42]

According to a former chief engineer at the Thruway Authority, Dewey chose the "worst possible place" to build a bridge.[43] The George Washington Bridge to the south is 4,760 feet long and the Bear Mountain Bridge to the north extends for 2,225 feet. A bridge at the Tappan Zee would be over 16,000 feet in length.

Not only would the bridge have to be long, but only a small portion of it would be supported by solid rock. At the Tappan Zee, rock is about seven hundred feet below the riverbed, and in the 1950s a bridge could not be built with footings that reached that level. Most of the three-mile-long structure would sit on top of soil, shells, and decaying timber, which themselves rest on sand, gravel, and clay.[44]

Given the soil conditions, the bridge's designers had to be creative. The center span would be supported by eight concrete boxes (caissons) that lie on the bottom of the river; the largest of these is half the size of a city block. The bridge's designer, Emil Praeger, had helped develop floating caissons during World War II when the Allied forces needed to create and protect portable harbors for the 1944 invasion of Normandy.[45]

The bridge was designed in a way that would minimize its weight and keep costs down; accommodating Korean War steel shortages would make it even lighter. In 1953, when no construction firms could afford to build the center span within the Thruway Authority's estimated budget, the authority modified the bridge's design to use even less steel.[46] The bridge also was built with a relatively thin concrete deck.[47]

Limitations imposed by the soil conditions would make the bridge less robust and subject to greater deterioration. The original designers introduced two hundred gaps (joints) in the deck so the bridge could accommodate settlement of the soft soils beneath the river. These gaps would later provide a direct route for deicing salts to leak onto the steel below. In addition, the bridge's structural support consisted of steel sections with holes designed to reduce weight and save steel costs. These holes also allowed deicing salt to penetrate inside the steel structure.[48]

The bridge's design had another flaw as well. The water from the deck was not channeled in drainage pipes to flow below the steel; the designers did not realize how much salt the Thruway Authority would use to melt ice. As a result, even more water, along with salt and debris, would flow directly from the deck and rust the steel below.[49]

Dewey's focus on short-term consequences explains the state's decision to build a bridge on the cheap. His choices reverberated for decades in the New York metropolitan area, since the problematic location and design would require costly bridge repairs in later years. Furthermore, many community leaders and local officials living in Rockland County near the Tappan Zee Bridge still have a lingering distrust of the Thruway Authority because of its secretive decision-making process in 1950.

In 1955, at a cost of $80.8 million, the New York State Thruway Authority opened the Tappan Zee Bridge (see figure 1.3). Because of the compromises that took place during its planning process, it was described decades later by a former Thruway Authority chief engineer as "a fragile structure that's been rotting away since it was built."[50]

Five years after the bridge opened, the New York State Department of Public Works (the predecessor of the New York State Department of Transportation) completed a new east-west highway, the eleven-mile Cross Westchester

FIGURE 1.3 Photo of the Tappan Zee Bridge with the New York City skyline in the background. The Tappan Zee Bridge is more than three miles long while the George Washington Bridge is less than one mile long. (*Source*: Neal Boenzi / *The New York Times* / Redux)

Expressway, to connect the New York State Thruway with I-95 (see figure 1.4). Except for a few property owners, there was little opposition to building this highway. Sy Schulman, who started working at Westchester County's planning department in the 1950s and later became the planning commissioner, remembers that no one seriously questioned the need for a major highway across the county.[51] The east-west portion of the Thruway in Rockland and Westchester Counties along with the Cross Westchester Expressway would be designated as I-287.

I-287 Fuels Suburban Growth

In 1950 Governor Dewey remarked, "If I were in the real estate business, I'd plunge on buying land in Rockland." The Thruway Authority's engineers predicted that Rockland County, previously a sparsely developed area, would be "greatly improved" by building the Thruway and new bridge.[52]

In 1954 Dewey told a joint session of the state legislature that the state's economy was largely based on highway transportation and that "the almost unbelievable economic benefits of the Thruway have already become evident where segments are in use. New industrial plants are springing up nearby. Residential areas are developing, particularly around the interchanges."[53]

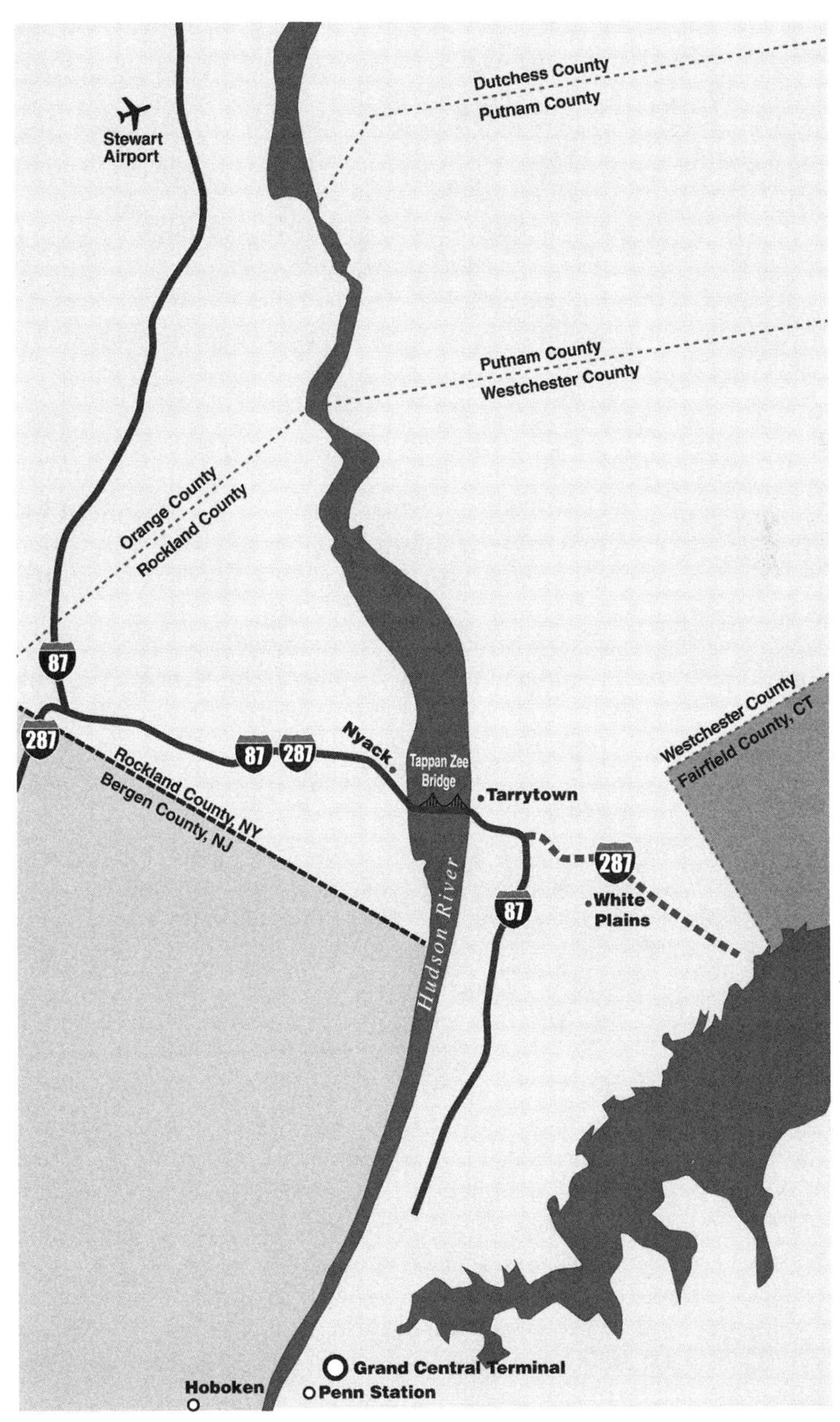

FIGURE 1.4 This map shows the Cross Westchester Expressway and the portion of the New York State Thruway that constitute I-287.

Dewey and the engineers were right. Once the Tappan Zee Bridge opened, rapid growth followed in the suburban regions traversed by the Thruway and I-287. Rockland's population increased by more than 20 percent in the four years following the opening of the Tappan Zee Bridge and by approximately 50 percent within a decade. Between 1950 and 1970, Rockland's population rose from 89,000 to 230,000.[54]

In 1950, Rockland had 408 farms with 17,376 acres in production. By 1969, the number of farms had dropped to 50 with only 4,022 acres in production. In 1993, ground was broken along I-287 for the Palisades Center, one of the nation's largest shopping malls. By 1998, only 344 acres of agricultural land were left in the county.[55]

Across the river, Westchester County also was transformed, especially after the Cross Westchester Expressway opened in 1960. Schulman remembers that none of Westchester's planners expected the commercial development that would occur along its path.[56] They had experience with the parkways that had been built decades earlier, but the Thruway and Cross Westchester Expressway were wider, carried commercial vehicles, and connected with the new interstate highway system.

The corridor was an ideal location for corporate headquarters because of the Cross Westchester Expressway's numerous entrance and exits, connections with the interstate system, proximity to Manhattan, and easy access from desirable residential neighborhoods. Prominent businesses such as AMF, General Foods, IBM, Nestle, ITT, and Texaco moved their offices to a portion of the corridor in the White Plains area known as the Platinum Mile.[57]

County officials, developers, and business leaders used their personal connections to entice New York City businesses to Westchester. New York City economic officials found that companies leaving the city were all moving in the direction of chief executives' homes.[58] Schulman said, "Of course, there's some truth that if you wanted to find out where a corporation was going to move, just look at where the chairman's wife shopped or who was on the board of golf courses."[59]

Between 1963 and 1971, a third of all new nonresidential construction in Westchester County took place within one and a half miles of the Cross Westchester Expressway, even though the area had only 8 percent of all the land in the county.[60] The amount of nonresidential floor space along the expressway grew rapidly: thirty-three million square feet in 1963, forty-eight million in 1972, sixty-two million in 1980, and sixty-nine million in 1984. In the early 1980s, the county expected the corridor to have ninety-four million square feet of office space with 170,000 employees by the turn of the century. Westchester was no longer just a New York City suburb but also a focal point for people to work and shop.[61]

In the 1980s, new types of suburban office and shopping centers, referred to as "edge cities" by author Joel Garreau, were growing across the nation.[62] Garreau cited three of them—Tarrytown, White Plains, and Purchase—along the I-287 corridor in Westchester. New Jersey had several edge cities along I-287 as well. In Connecticut, Stamford was another edge city; it was only a ten-minute drive from the I-287 corridor.

South, west, north, and east of Manhattan, the I-287 corridor had become an important circumferential highway and business corridor. There was a dark side to this growth, however.

Since the end of World War II, the American Dream has been defined as a house in the suburbs. The Sierra Club explains how the dark side of explosive suburban growth has been increased traffic congestion, environmental degradation, and dying city centers.[63]

In the 1970s, urban job losses led to a fiscal crisis in New York City. The Bronx, the city's borough adjacent to Westchester County, lost 21 percent of its population in the 1970s. Westchester's economy was no longer centered in the cities of Yonkers and Mount Vernon near the New York City border, but rather along I-287, which housed one-third of Westchester's total commercial office space by the end of the 1980s.[64]

The suburban development in Rockland and Westchester was not unique. In the 1950s and 1960s, many U.S. metropolitan areas experienced a mass migration of white middle-class families from older industrial cities to their suburbs.[65] In America's new suburbs, almost everyone lived beyond walking distance of stores, schools, churches, and public transportation services.[66] In previous generations, Westchester and Rockland residents had relied on trolley cars, railroads, ferries, and their feet to take them to central business districts. In the new suburban communities, they drove on highways to shopping centers and office parks.

Long before the I-287 corridor existed, Westchester County was one of the wealthiest counties in the nation. Bankers and corporate executives commuted by train to New York City's Grand Central Terminal from the affluent suburbs. In the early twentieth century, these communities had deliberately used their zoning and other regulatory powers to attract "class" rather than "mass" from New York City.[67]

After I-287 was built, the demand for affordable housing by Manhattan and Westchester workers pushed new residential development to the edges of the New York metropolitan area. At the same time, Westchester towns and cities encouraged new office development because it increased tax revenues without adding to school costs.

The region's new residential and commercial development along I-287 exacerbated sprawl and led to increased traffic congestion. By the late 1970s, traffic volume on the Tappan Zee Bridge had grown so dramatically that serious congestion was occurring nearly every weekday, eastbound in the morning and westbound in the evening.[68] At the end of a long summer weekend, a six-mile backup at the Tappan Zee Bridge's tollbooths was not unusual.[69]

During the bridge's first year of operation, 18,000 vehicles crossed the span every day. That increased to 29,000 in 1960; 66,000 in 1970; and 74,000 in 1980. In 1980, the state projected that traffic volumes would keep increasing at a rapid rate, and that turned out to be accurate. Traffic volumes reached 78,000 in 1982; 81,000 in 1983; 112,000 in 1990; and 135,000 in 2010.[70]

Not only had traffic grown faster than expected, but cars were going in unexpected directions. The Thruway's designers assumed that most drivers traveling east over the Tappan Zee Bridge would continue south on the Thruway toward New York City. However, it turned out that most drivers crossing the bridge in 1980 wanted to get onto the Cross Westchester Expressway, regularly causing major backups.

In the early 1980s, Westchester County's business community became increasingly concerned that traffic congestion would adversely affect its economic vitality. A coalition of construction firms warned that the county was facing a crisis situation along I-287 and that unabated traffic congestion could halt all new development.[71]

By 1982, Schulman was president of the Westchester County Association, the county's leading business group. He told state officials that if workers outside the county could not get to jobs in Westchester, employers and developers would go somewhere else.[72] Despite the negative aspects of the suburban development, the state wanted to accommodate even more business and residential growth along the corridor. Adding new highway capacity had gotten much more complicated, however.

In the 1950s and 1960s, the state built numerous bridges and highways in the New York City area, including the Long Island Expressway and the Verrazano-Narrows Bridge. However, by the 1970s, the negative effects of highways on the environment and urban areas had become well known. Passage of the National Environmental Policy Act of 1969 required states to conduct public hearings and document the environmental impacts of new federally funded roads and bridges. In the early 1970s, environmental concerns led to the cancellation of several transportation projects in New York, including a new highway along the western edge of Westchester and a new bridge between the eastern end of Westchester and Long Island.

Regional Efforts to Tackle Congestion

New York considered adding capacity to the Tappan Zee Bridge in the 1970s and 1980s, but that idea ultimately was deemed too expensive and disruptive, even though both of the major Hudson River bridges north and south of the Tappan Zee Bridge have been expanded to accommodate increases in vehicle traffic. The George Washington Bridge, seventeen miles south of the Tappan Zee, connects New Jersey and New York City. It opened in 1931, and the Port Authority added a second deck in 1962. Two decades later, Governor Hugh

Carey's senior transportation advisor asked the Thruway Authority about the possibility of double-decking the Tappan Zee Bridge, as well. Its executive director laughed and said "it would fall in the river."[73]

The Newburgh-Beacon Bridge crosses the Hudson River thirty-five miles north of the Tappan Zee. It was completed in 1963, and in the early 1970s the state decided to build a new bridge parallel to the existing one.[74] Building the new span between Newburgh and Beacon was relatively inexpensive and uncontroversial compared to widening the Tappan Zee Bridge. The Newburgh-Beacon Bridge is half as long as the Tappan Zee Bridge and the approach roads in Newburgh and Beacon were already wide enough to accommodate a second span.[75]

In 1973, the Thruway Authority told the New York State Department of Transportation (NYS DOT) that it needed an additional six lanes across Rockland County and over the Hudson River to the Cross Westchester Expressway because the eighteen-year-old bridge had become overburdened. Three years later, Thruway Authority chairman Gerald Cummins was dismayed that the state still was not doing anything. He wrote that the failure to expand the existing Thruway "would have a devastating effect with attendant horrendous congestion." Cummins claimed widening the Thruway from six to twelve lanes would have minimal environmental affect. He noted only a few residential properties would have to be taken and no schools, churches, or industries would have to be relocated.[76]

The Thruway Authority was concerned that a flood of vehicles would try to cross the Tappan Zee Bridge after the New Jersey Department of Transportation (DOT) completed a connection between the Thruway and its portion of I-287. That interchange would make it easier for truckers traveling to and from New England to avoid the river crossings in New York City. However, no one knew exactly how many more cars and trucks would drive over the Tappan Zee Bridge after the connection was completed. The NYS Thruway Authority, NYS DOT, and the New Jersey DOT each developed their own conflicting estimates.[77]

In 1978, the Thruway Authority still thought another span was needed to accommodate future traffic. The chief engineer said, "There's no question we're up to all it can take." But a new bridge was too expensive for the authority to build on its own. The Thruway Authority did not have the ability to issue enough bonds, and toll road operators at the time were not eligible to receive federal funds.[78] Furthermore, state law did not allow the Thruway Authority to acquire any real property by eminent domain on its own; it had to do so through NYS DOT.[79]

At least two interest groups were hoping for a new bridge. In 1978, after Ross Pepe started a trade organization to represent construction firms in the Hudson Valley, he told NYS DOT commissioner William Hennessy that a new bridge would generate new and vital development that would have a positive impact on the state's economy.[80] Pepe's organization set up a bridge study committee and argued that if the growing traffic were not accommodated, the region would face serious problems. The Automobile Club of New York also wanted a new bridge but did not expect one to get built.[81]

NYS DOT was responsible for the Cross Westchester Expressway and more than fifteen thousand miles of other free highways and bridges across the state. Its officials were not expecting to see a new bridge either, given the high cost and environmental concerns. The original Tappan Zee Bridge had cost $81 million in 1955. By the late 1970s, NYS DOT estimated that widening the existing bridge would cost between $500 million and $1 billion. One NYS DOT official said, "There is no place to put it in Westchester" and "the environmental impact would be horrendous."[82] Another NYS DOT official commented, "We may end up with a problem, but maybe it will be better to live with the problem than have a solution."[83]

In 1980, Westchester County Association president Sy Schulman commissioned Richard Newhouse of the Automobile Club of New York to conduct a traffic study. Schulman soon realized that a new bridge would be politically and financially infeasible. He figured that carpooling and vanpooling would be the only source of relief available to Westchester employers and employees, although that was not likely to reduce traffic by more than 15 percent.[84]

Yonkers Mayor Angelo Martinelli thought the best place to build a new bridge would be about nine miles south of the Tappan Zee, between his struggling city and New Jersey. NYS DOT engineers showed him the two-hundred-foot right-of-way they would need to build a new bridge. A connection to the downtown business district would require condemning even more properties. Martinelli changed his mind when he realized the new bridge would destroy rather than revitalize the city.[85]

Traffic along I-287 was symptomatic of a traffic problem that was much bigger than just one corridor. New York officials were dealing with a region-wide problem that affected businesses and residents in New York, New Jersey, and Connecticut. The three states had actually set up a tristate planning group to consider regional transportation issues in the 1960s. That organization grew into the Tri-State Regional Planning Commission, which was given responsibility for reviewing and approving all requests for federal transportation, housing, and land-use funds in a seventy-five-mile radius from Times Square. The commission had 220 employees with offices on the eighty-second floor of the World Trade Center. Its jurisdiction covered eighteen million people, including two-thirds of New York State's population, and encompassed 586 local governments.[86]

However, the Regional Planning Commission's agenda was largely set by the states, and it made few independent recommendations. It maintained a low profile by focusing on technical analyses of growth patterns and transportation demands.[87] When the commission did try to perform an independent planning role it met resistance, since region-wide solutions inevitably contradicted the efforts of a jurisdiction.[88] In 1978, New York's governor, Hugh Carey, removed two suburban counties on Long Island from the commission's jurisdiction after they made a fuss about a commission report addressing the needs of low-income families.[89] Two years later, the Regional Planning Commission was dissolved after Connecticut withdrew and New Jersey was not interested in being part of a bistate planning organization.[90]

Out of the ashes of Regional Planning Commission, the State of New York created the New York Metropolitan Transportation Council, whose jurisdiction would be limited to programming federal transportation funds for New York City and Long Island, as well as Westchester, Rockland, and Putnam Counties. But the Metropolitan Transportation Council was given limited powers relating to the I-287 corridor. The municipalities retained control of land-use decisions while most of the major transportation decisions about the I-287 corridor still would be made by NYS DOT and the quasi-independent NYS Thruway Authority.

In 1980, New York officials began a quixotic search to solving the corridor's congestion problem. Just as there are no easy ways to build a new bridge in a densely populated urban area, there are no practical and effective solutions to eliminating traffic congestion.

The highly regarded author Anthony Downs explains in his book *Still Stuck in Traffic* that once peak-hour congestion has appeared on a major roadway, there is no way for a region to build its way out of it. There are just too many people seeking to use the roads at the same time each day for this approach to work without enormous financial and environmental costs. He writes, "Key routes would have to be widened so much that huge portions of the entire region would be turned into giant concrete slabs."[91] Building more lanes would encourage people who traveled on different roads or at different times to take advantage of the new capacity. It would also encourage more development, which would only worsen the congestion problem.

Downs finds that the alternatives to adding highway capacity are no more appealing. Charging tolls on all major roads is politically unacceptable and providing enough public transit is impractical given America's suburban auto-oriented patterns of development. Even though the I-287 congestion problem could not be solved, there would be no shortage of people peddling their solutions over the next three decades.

2

Searching for Congestion Solutions (1980–1988)

By the early 1980s, Westchester County business leaders and elected officials became increasingly alarmed that traffic along I-287 would become intolerable, weakening the county's economy and impeding its growth. Officials considered a wide range of solutions, but settling on a plan that the major stakeholders could agree upon proved elusive.

Reaching a consensus was difficult for several reasons. Officials at the two separate state entities responsible for I-287 had clashing strategic interests. The New York State Department of Transportation (NYS DOT) wanted to reduce the number of people driving alone, while the New York State Thruway Authority wanted to increase toll revenues and please its bondholders. Transportation officials also had to deal with the competing opinions and interests of various governmental, planning, and citizen groups. Underlying their difficulty was the fact that the state agencies were working under a great deal of uncertainty. Although they had sophisticated staff and state-of-the-art analytical tools, no one really knew how many people would actually drive, share rides, and ride buses across the corridor under various scenarios.

The Studies to Reduce Congestion Begin

In the late 1970s, commuters and elected officials frequently called and wrote letters to NYS DOT to express their frustration with I-287 traffic. Newspapers regularly reported about traffic delays on the Tappan Zee Bridge and at the

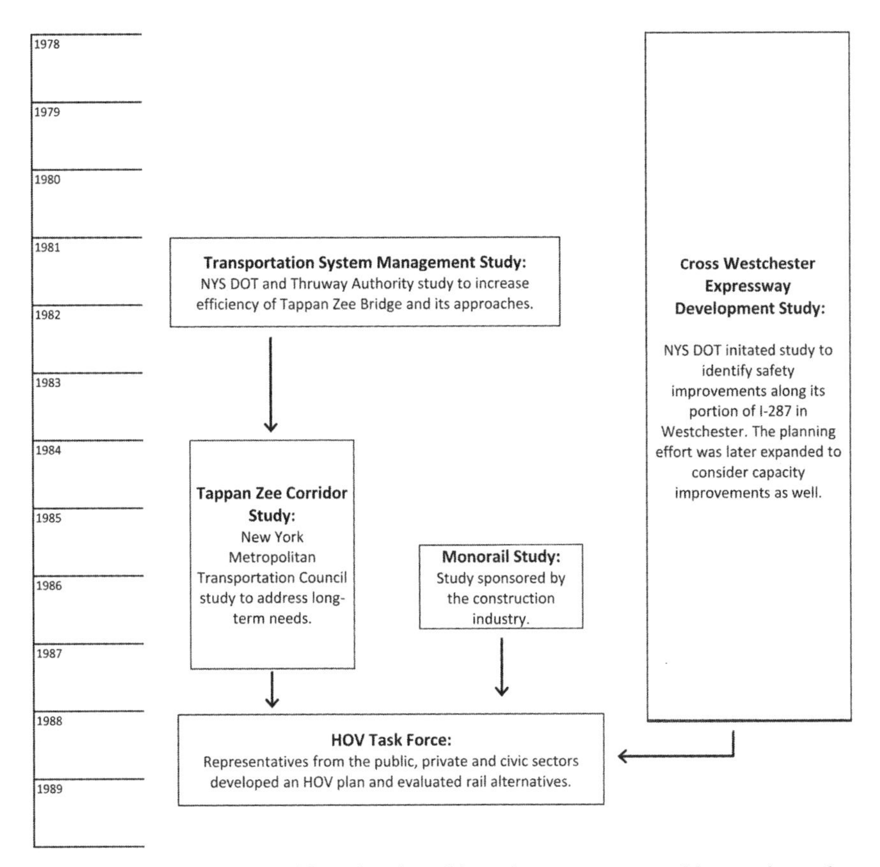

FIGURE 2.1 In the 1980s, several formal studies addressed transportation problems and considered potential improvements along the I-287/Tappan Zee Bridge corridor.

interchange between the Thruway and the Cross Westchester Expressway.[1] This set in motion a series of formal studies starting with the Transportation System Management Study and the Cross Westchester Expressway Development Study (see figure 2.1).

In 1980, Richard Peters, an engineer in NYS DOT's regional office in Poughkeepsie, started working with Thruway Authority engineer Duane Dodds to identify potential solutions to the traffic congestion along the I-287/Tappan Zee Bridge corridor.[2] The Poughkeepsie office was responsible for the state's highways in New York City's northern suburbs, including the Cross Westchester Expressway portion of I-287, while the NYS Thruway Authority was responsible for the Thruway portion of I-287, including the Tappan Zee Bridge.

Peters hoped to simultaneously reduce congestion and pollution, conserve energy, and avoid the need to build a new bridge. Although NYS DOT had long been associated with building new roads and widening existing ones, fiscal constraints, environmental awareness, and the energy crises of 1973 and 1979 had led to an emphasis on new ways of moving people. Highway engineers were beginning to explore ways to increase the capacity of highways without building new lanes and roads.

In February 1981, Peters helped draft a proposal to the U.S. Department of Transportation seeking $500,000 in transportation system management funds for the I-287 corridor.[3] He remembers that the concept of transportation system management, which consists of relatively low-cost measures designed to increase the efficiency of existing highway and transit facilities, was the "hot thing" for transportation planners in 1981.[4]

Peters realized that the I-287 congestion resulted from traffic demand exceeding capacity during the peak travel time periods. His 1981 proposal suggested both adding capacity and reducing demand. The Thruway Authority could add capacity by converting the Tappan Zee Bridge's existing ten-foot-wide center median into an additional travel lane. This would in effect convert a six-lane bridge into one with seven lanes (see figure 2.2). The new lane could be designated as a high-occupancy vehicle (HOV) lane, which would be limited to cars with two or more persons.

Peters proposed to reduce demand by lowering bridge tolls for carpoolers, establishing ride-share programs, starting shuttle bus services, and building park-and-ride lots. Local employers could help reduce demand by limiting parking spaces at their office parks and offering modified-work-hour plans for their employees. These demand-reduction measures would increase the average number of persons in each vehicle and spread traffic over a longer peak period.

The state's proposal indicated that the Thruway Authority could charge higher tolls during peak periods, a strategy that would later be known as *variable pricing* or *congestion pricing*. Although NYS DOT identified a wide of range of alternatives, Peters suspected it would be next to impossible for the state to implement some of them, such as limiting free parking at corporate office parks.[5]

Although the federal government turned down the state's proposal, representatives of the Thruway Authority and NYS DOT subsequently met in August 1981 to discuss possible next steps. The Thruway Authority's deputy executive director, Daniel Garvey, said the two state transportation organizations could conclude their efforts by indicating no improvements were necessary since traffic conditions on the Tappan Zee Bridge were similar to conditions on the tunnels and bridges heading into New York City.[6] NYS DOT proposed they complete the study anyway. The two transportation agencies subsequently agreed to undertake what became known as the Transportation System Management Study for the Tappan Zee Bridge and Its Approaches.

During the Transportation System Management Study, Rich Peters worked with Thruway Authority engineers to reevaluate the ideas they generated in their application for federal funds. They recommended converting the bridge's

Six lanes with median

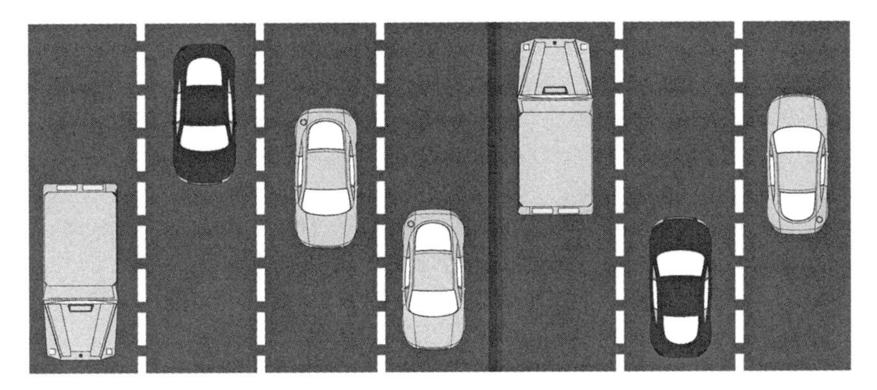

Four eastbound and three westbound lanes

FIGURE 2.2 Converting the Tappan Zee Bridge's median strip would create a total of seven lanes. (*Source:* NYS DOT, Metro-North, and NYS Thruway Authority, "Tappan-Zee Bridge/ I-287 Corridor Project," Status Briefing to County Executive R. Astorino, March 5, 2010)

median into a new eastbound lane and taking actions to reduce demand, such as building park-and-ride lots and dedicating the new bridge lane to high-occupancy vehicles.[7] They also suggested using toll pricing to encourage higher vehicle occupancy, although Thruway Authority officials made sure the study's final report noted that the authority was prohibited by its bond covenants from adopting a policy that would produce a deliberate loss in revenues.

After releasing the study in 1982, NYS DOT officials told the *New York Times* that the idea of a second bridge was not feasible, given a price tag between $750 million and $1 billion and stiff opposition from both sides of the river. The officials did say, though, that in fifteen or twenty years, the state might have to reconsider such a plan.[8]

At two public information meetings held to discuss the study's recommendations, the Thruway Authority and NYS DOT officials emphasized the report's capacity-increasing aspects, most specifically converting the median into a new travel lane.[9] They claimed that adding a seventh lane would be more effective than measures to reduce peak period demand. Congressman Benjamin Gilman said the need for a second bridge was inevitable. State Senator Linda Winikow argued that "Rockland was born out of the bridge, and we are going to die unless we find some alternative to the current traffic problems."[10]

Many of the meetings' attendees thought the state's ideas did not go far enough. Westchester County Association President Sy Schulman (Westchester County's former chief planner) expressed his support for carpooling, vanpooling, and flexible hours for employees.[11] The Yonkers mayor, Angelo Martinelli, no longer asked for a new bridge; now he wanted the state to pay for new ferry services, instead.[12] The idea of a twenty-three-mile-long monorail line also was raised.[13] Given its high cost and the suburban nature of the corridor, the idea seemed absurd to transportation engineers—but not to the public.

A few weeks after the public meetings, NYS DOT and Thruway Authority staff sat down to talk about the next steps they would take to reduce congestion. The two agencies concluded that the ferry service and monorail ideas raised by the public were not feasible given the dispersed travel patterns in the area. They also agreed the state should conduct a long-term study to determine the actual environmental impacts of a new bridge and to "zero in on the cost."[14]

According to a memo Peters wrote after the meeting, the two agencies agreed that the Thruway Authority would convert the median into a seventh lane, and that it would be dedicated for high-occupancy vehicles if the state did not build a new bridge.[15] The Thruway Authority went ahead and started working on converting the six-lane bridge into a bridge with four eastbound and three westbound lanes, a $30 million project that was completed in 1987. The seventh lane would be used for eastbound traffic because tolls were collected for vehicles traveling in that direction.

The Thruway Authority and NY DOT agreed to consider the long-range needs of the transportation corridor in yet another study, which became known as the Tappan Zee Corridor Study. Both agencies initially expected the agencies to focus on a new bridge over the Hudson River, but other stakeholders had a broader perspective. In March 1983, Schulman said the state should undertake measures to reduce single-occupancy vehicles, since the new lane on the bridge would provide some relief but offer no flexibility for future growth. NYS DOT commissioner William Hennessy agreed to make HOV lanes an important component of the study.[16]

Two months later, the study's goals were made even broader after the head of the Office of Public Transportation in NYS DOT's Albany headquarters recommended to the commissioner that the study include a full range of modal alternatives such as ferries, hydrofoils, a monorail, a new bridge, and toll pricing strategies.[17] NYS DOT officials expected that their work would take two years. Instead, it took four.

To conduct the study, the NYS DOT turned to the staff of the New York Metropolitan Transportation Council, which was responsible for programming all federal transportation funds in New York City, Long Island, Rockland, and Westchester. From NYS DOT's perspective, the council had available staff and technical expertise.[18] The council would also consider a larger geographical area than the jurisdiction under NYS DOT's Lower Hudson Valley regional office.

The Tappan Zee Corridor Study commenced in 1984 and was guided by representatives from NYS DOT, Thruway Authority, Westchester County, and Rockland County. Using new computer mainframe software designed to estimate future traffic conditions under various scenarios, the council's planners on the eighty-second floor of the World Trade Center found that severe traffic congestion on the Tappan Zee Bridge would occur daily by 2010 during both morning and evening peak periods.[19]

The planners considered potential locations for a new bridge and estimated that construction costs would range from $510 million (for a bridge adjacent to the existing one) to $1.27 billion (for a new bridge between New Jersey and New York). They evaluated other options as well, including new ferry services across the Hudson River, new express bus services, light rail, and monorail. They also considered bringing back passenger service to the West Shore Line, a rail freight line that runs along the west shore of the Hudson River.

In 1985, the planners reported that the most cost-effective strategies were not the big-budget alternatives, but measures that would shift people out of their single-occupant vehicles during peak periods. This type of action was very similar to the transportation system management strategies considered by NYS DOT and the Thruway Authority a few years earlier but now were known as *travel demand* management (or transportation demand management) measures.

The strategies recommended by the council's planners included new park-and-ride lots, bus services, employee-sponsored vanpools, and variable work hours. They determined that one of the most effective measures to reduce peak period demand would be to charge higher tolls during the peak period on the bridge.

A survey revealed that 82 percent of Tappan Zee Bridge travelers between 7:00 A.M. and 9:00 A.M. were making their entire trip in single-occupant vehicles. Commuters who regularly crossed the bridge were given a 50 percent discount; they paid seventy-five cents instead of the normal $1.50 toll.[20] The computer models showed that if the Thruway Authority simply eliminated the discount during this period, peak-hour volumes on the bridge would go down by as much as 10 percent. Most of those travelers would continue using the bridge, but they would cross it before or after the beginning of the peak period.[21] Eliminating the discount would also generate $2.5 million a year in additional revenue.

Although the planners found that increasing tolls showed great promise in reducing peak period congestion, the Thruway Authority was boxed in. Raising tolls for single-occupant vehicles was financially beneficial but politically difficult, while lowering tolls for carpoolers was financially detrimental but politically popular. The Thruway Authority did not talk about its dilemma; rather, the authority reiterated that it was prohibited by its bond covenants from adopting a policy that would produce a deliberate loss in revenues. NYS DOT officials felt that the Thruway Authority was repeatedly using that as an excuse not to change its policies.[22]

The transportation council's planners realized that the Thruway Authority had a greater opportunity to reduce peak period congestion on the Tappan Zee Bridge than the Port Authority did on New York City's three Hudson River crossings. The Tappan Zee Bridge was just as heavily congested during the peak periods as the city's crossings, but during the off-peak periods the Tappan Zee saw considerably less traffic than the George Washington Bridge, Lincoln Tunnel, and Holland Tunnel.[23] That meant the Thruway Authority could more readily spread out peak period demand.

Overall, the council's planners estimated that their measures would reduce the number of vehicles during the peak period by 18 percent. The planners found that an HOV lane would not make much of a dent in the congestion problem: spending hundreds of millions of dollars for an HOV lane would only reduce the number of vehicles by an additional 2 percent. Although some cars would shift from the general-purpose lanes to an HOV lane, the general-purpose lanes would soon fill up again.

In a 1985 interim report, the planners concluded that building a new HOV lane would not draw a sufficient volume of high-occupancy vehicles to justify establishing such a facility. It would entice few additional people to shift from cars to public transportation because potential bus riders were found to be more influenced by service frequency and fare levels than by travel-time savings.[24]

Just as in the previous analyses, the planners found that new rail and ferry services would not significantly reduce congestion levels and were not cost-effective. The final report indicated that future growth might require a new Hudson River crossing, and that the best location for that would be adjacent to the Tappan Zee Bridge because it would make the most effective use of existing highways. The study's May 1987 final report recommended measures to encourage the use of carpooling, vanpooling, and buses. It also suggested adjusting bridge tolls to encourage higher vehicle occupancy.

Comparing the study's 1985 interim report issued by the council with the 1987 final report reveals a glaring change relating to the HOV lanes. In 1985, the council's planners deemed HOV lanes to be ineffective, but the 1987 final report suggested that the state build an HOV lane if it gained broad support.[25]

What happened? The planner's findings about the lack of an HOV lane's effectiveness were overruled by Poughkeepsie and Albany officials because NYS DOT had initiated yet another study—the Cross Westchester Expressway Development Study. That effort, which focused on NYS DOT's portion of I-287, concluded that an I-287 HOV lane should be built on the expressway, and that its success required an HOV lane and carpooling toll discounts on the Tappan Zee Bridge.

HOV Lane Saga Begins

In 1978, well before the commencement of the Transportation System Management and the Tappan Zee Corridor studies, NYS DOT's Poughkeepsie office began the Cross Westchester Expressway Development Study to consider the installation of a concrete median barrier and other safety improvements on the expressway. This analysis later expanded to evaluate not only safety improvements but also rehabilitation of the entire expressway.

The expressway had an accident rate substantially over the average for the state. The greatest number of incidents occurred at interchanges, which because of their close proximity to each other generated significant volumes of merging and weaving traffic.[26] Although federal guidelines recommended that interchanges on interstate highways be spaced at least one mile apart, the Cross Westchester Expressway had fourteen along its eight-mile length, some of which had several entrances and exits.[27]

NYS DOT had other reasons to rehabilitate the expressway as well. Motorists were understandably concerned after they heard that a truck driver had carelessly driven over the short median fence into traffic heading in the opposite direction; in addition, many people complained about the noise resulting from trucks going over deteriorated portions of the pavement.[28] Furthermore, NYS DOT wanted to address the concerns of business leaders, motorists, and elected officials who were complaining about traffic, which was flowing at less than thirty miles per hour, stop and go, for about one and a half hours during the morning peak periods. NYS DOT reported that the "expressway must be able to deliver the maximum number of employees to Westchester's office building with the least possible amount of delays."[29]

Since rehabilitating the expressway to address safety concerns would inconvenience and aggravate commuters, the engineers thought that "drivers should get something for it," and they assumed that everyone would want a wider highway.[30] To continue carrying present and future traffic volumes, NYS DOT's regional office decided to widen the six-lane expressway to eight lanes at an estimated cost of $66 million. In 1983, the regional office considered the project to be a minor rehabilitation and expected that work could be completed in 1988.[31]

While the planners working on the Transportation System Management and Tappan Zee Corridor studies were considering ways to reduce the number of vehicles along I-287, their engineering counterparts were recommending a more traditional way of dealing with traffic: buy more property, knock down more trees and homes, and build more lanes.

NYS DOT's engineers working on the Cross Westchester Expressway Development Study considered building an HOV lane on the expressway but decided that since there was no such lane on the Tappan Zee Bridge, it would not provide sufficient travel time savings to induce enough people to carpool and use transit. In 1983, they concluded that if the Tappan Zee Bridge were to ever get an HOV lane, then they could build one along the median of an eight-lane expressway.[32] Figures 2.3, 2.4, 2.5, and 2.6 show four options considered by NYS DOT in the 1980s and early 1990s.

Selecting one of the four options was not a trivial decision. Improving the highway could cost hundreds of millions of dollars, affect the travel patterns of hundreds of thousands of people, influence where people would decide to live and work, and increase traffic on connecting and parallel roadways.

In the mid-1980s, NYS DOT shifted back and forth between recommending an eight-lane-wide expressway or a six-lane expressway with a reversible

FIGURES 2.3–2.6 Studies in the 1980s considered four options, among others, to improve safety and lessen congestion along the Cross Westchester Expressway. Fig. 2.3 (*upper left*): This image shows the existing configuration along the Cross Westchester Expressway. Fig. 2.4 (*upper right*): This image shows a proposed eight-lane configuration along the Cross Westchester Expressway. Fig. 2.5 (*lower left*): This image shows a proposed six-lane configuration along the Cross Westchester Expressway. Fig. 2.6 (*lower right*): This image shows a proposed reversible HOV lane along the Cross Westchester Expressway. (*Source:* Federal Highway Administration and NYS DOT, *I-287/Cross Westchester Expressway: Final Design Report/Final Environmental Impact Statement*, June 1997)

HOV lane. Engineers and planners considered criteria such as cost, property takings, and the ability to accommodate future growth. But there was no clear best answer. To complicate matters, NYS DOT officials were getting mixed signals from the Thruway Authority about whether the authority would designate their new seventh lane as a reversible HOV lane. The Thruway Authority engineers were cooperative but expressed their concerns that not enough people would use the HOV lane to make it successful.[33]

The disagreement on whether the seventh bridge lane was to be dedicated to high-occupancy vehicles exposed a fundamental difference in the perspectives and goals of the people representing the two transportation organizations. While NYS DOT's planners wanted to maximize the number of commuters who could cross the bridge, the Thruway Authority's officials had bonds to pay off and faced political pressure to keep tolls low. The very first report that the Thruway Authority's executive director received in the morning was the previous day's toll revenue.[34] A congested general-purpose lane would be much more lucrative to the Thruway Authority than one that was designated for high-occupancy vehicles. The Thruway Authority's revenues would be further constrained if they took NYS DOT's suggestion and offered carpoolers a discount on a new HOV lane.

NYS DOT also was receiving conflicting public comments. Communities along the expressway were concerned about the additional noise, traffic, and property takings associated with an eight-lane roadway. They preferred keeping the expressway at six lanes. However, the City of White Plains, Westchester's largest municipality along the corridor, supported the eight-lane option.

Likewise, Andrew O'Rourke, the Westchester County executive, supported eight lanes and was skeptical about the effectiveness of carpooling.[35] As he told the *New York Times*, "The hardest thing in Westchester County is to get people out of their cars. It's like separating John Wayne from his horse."[36]

The HOV alternative got a boost in May 1985 when Governor Mario Cuomo appointed Franklin E. White to be the new commissioner of NYS DOT. White had been Virginia's secretary of transportation, and he was familiar with a successful HOV lane in northern Virginia.[37] He saw HOV lanes as having a promising future and believed that a successful implementation of one along I-287 would save the state billions of dollars by obviating the need to construct a second Tappan Zee Bridge or a ten-lane Cross Westchester Expressway. Although no northeastern state had ever before widened a highway to build an HOV lane, White appreciated its potential to improve traffic and realized that building one would give NYS DOT positive national attention.[38]

After White's appointment, NYS DOT officials in Albany began extolling the benefits of HOV lanes and continued urging the Thruway Authority to designate one on the Tappan Zee Bridge's seventh lane when it opened.[39]

By August 1985, Rich Peters and the head of the NYS DOT regional office's planning group, Alan Bloom, had an intuitive sense that travel demand strategies alone would not encourage enough drivers out of their cars and into carpools and buses. They thought drivers needed to see that carpooling and transit could save them time as well as money.[40] The next year, NYS DOT's consultants on the Cross Westchester Expressway study came to the same conclusion. Based on the consultants' experience, express bus service without an HOV facility of adequate length would not be successful, and an HOV lane without express bus service would not be justifiable.[41]

In 1986, NYS DOT's analysis revealed that only two alternatives (ten lanes and six lanes with an HOV lane) provided sufficient transportation capacity to accommodate the increased travel across the bridge that would stem from the twenty-five million square feet of nonresidential development and nearly sixty thousand new jobs expected along the corridor in the next fifteen years.[42] In 1987, NYS DOT took a stand in favor of the HOV lane. Commissioner White's support for HOV lanes was a critical factor in its decision.[43]

NYS DOT officials now wanted the results of the Cross Westchester Expressway Study to be consistent with the results of the New York Metropolitan Transportation Council's Tappan Zee Corridor Study. Although the council's planners had previously found the HOV lane to be ineffective, the study's final report in May 1987 declared that if there is "broad support by local government, corporate sector, and users," an HOV lane should be constructed from Rockland to Westchester.[44]

In June 1987, NYS DOT officials released the results of their Cross Westchester Expressway analysis.[45] They recommended the highway be reconstructed with six wider lanes along with an HOV lane. The HOV lane would be reversible, similar to the one that they were advocating for the Tappan Zee Bridge. It would be used eastbound in the morning and westbound in the evening. As part of the expressway enhancement, several interchanges would also be upgraded to address safety, weaving, and merging problems. NYS DOT reported that the expressway could accommodate future growth with this configuration because a reversible HOV lane had the potential to carry three times as many people as a general-purpose lane.

Trying to Generate Support

NYS DOT officials realized that strong citizen and official support would be necessary to implement the reversible HOV lane, start new express bus services, build new park-and-ride lots, and change tolls. Support from state legislators and the Westchester County executive would be particularly important. However, as the state was planning and promoting its concept, all the important institutions and elected officials in Westchester seemed to have their own ideas for addressing congestion, and none of them seemed to favor the state's HOV idea.[46]

The automobile club and construction industry wanted a wider highway. Two transit advocacy organizations wanted the state to build a light rail line, which they claimed would be needed to avoid gridlock in ten years.[47] A dozen communities along I-287 wanted to minimize noise and property takings; they were also were concerned about the additional traffic that would result if the HOV lane were ever converted into general-purpose use.[48]

A state assemblyman arranged a meeting between Westchester's state legislative delegation and NYS DOT in order to learn more about the state's planned transportation improvements.[49] NYS DOT officials told the assembled legislators that an HOV lane would provide enough capacity to the year 2020 yet would require few displacements and have minimal environmental impacts. They also stated that the project would improve pavement and interchanges for all users, add noise barriers, and install a concrete median barrier.[50] NYS DOT officials took pains to explain how the HOV lane was a flexible solution. It was compatible with either the existing Tappan Zee Bridge or a potential new parallel bridge, they pointed out. Also, it could be converted into two general-purpose lanes if it failed, and it did not preclude light rail or monorail.

The general consensus was that there should be some type of capacity improvements, and that an HOV lane should allow for conversion to general use lanes if it failed to effectively reduce congestion. The delegation also advocated further study of light rail and monorail solutions.[51]

A state assemblyman named George E. Pataki was a member of the delegation. Six months after the meeting, Pataki told NYS DOT Commissioner White that the single reversible high-occupancy lane in the median of I-287 would be completely inadequate to meet future needs. Given Westchester's rapid growth, he said there was no question in his mind that the two-lane HOV alternative must be adopted.[52] As a Republican member of the Democrat-controlled state assembly, Pataki had little influence. It was a different story, however, seven years later when Governor George E. Pataki was making all the decisions regarding I-287.

Obtaining support from Westchester County Executive Andrew O'Rourke was crucial to generating political support, obtaining funding, and ultimately ensuring the success of the HOV lane. The department had an unwritten policy that it would not advance a major project in a county whose county executive opposed it.[53]

O'Rourke's opinion affected state funding because he was influential among the county's state legislative delegation. He also had an important voice in federal funding decisions because he was a voting member of the New York Metropolitan Transportation Council. Since the council required unanimous votes before allocating the region's federal transportation funds, O'Rourke had a de facto veto over the project. Even after the HOV lane opened, O'Rourke had the power to make or break the project since the state needed the county to operate bus services and encourage ride-sharing programs.

Unfortunately for NYS DOT, O'Rourke did not support the HOV lane concept. He wanted the expressway expanded to eight lanes because he thought the HOV lane was impractical, needlessly expensive, and would have limited effectiveness in easing the corridor's congestion. He thought that if it were ever warranted in the future, the two new lanes of an eight-lane highway could be designated for HOV use.[54]

NYS DOT maintained that an eight-lane highway would only provide enough capacity until the year 2000, while it was more expensive and would require taking more homes than the HOV lane.[55] NYS DOT officials also were skeptical of Westchester's idea that general-purpose lanes would ever be converted into HOV lanes in the future, even if such an action was justified.[56] They believed (and were proven correct) that once general-purpose lanes were heavily used, dedicating them for HOV use would be nearly politically impossible.

By the end of 1987, O'Rourke felt sure that both the eight-lane highway and HOV lanes would only serve as interim measures to solve the congestion problem. In a letter to Commissioner White he wrote, "Each day I grow more convinced that the future of the Cross Westchester Expressway/Tappan Zee Bridge corridor lies largely in ambitious mass transit in and between Westchester and Rockland Counties, not in the construction of a second highway bridge across the Tappan Zee or in continual expansion of the Cross Westchester Expressway." He thought that the HOV lane was costly and cumbersome and would make the installation of a rail line more expensive and less attractive in the future.[57]

A few days after NYS DOT's meeting with the legislators in January 1988, O'Rourke went up to Albany to meet with White.[58] The commissioner was insistent that the HOV lane would provide adequate capacity beyond the year 2020. He explained to the county executive that if the HOV lane was unsuccessful, it could always be converted to two general-purpose reversible express lanes.[59] O'Rourke told White that although he preferred an eight-lane highway, if the state were to go ahead with the HOV lane, it should provide for one in each direction, not a single reversible lane. Given plans for new development along the corridor, O'Rourke expected that there would be demand for two-way peak-period HOV lanes in the future.

Although White was concerned about the additional cost and property takings that would be required with an even wider expressway, he promised the county executive that NYS DOT would study the potential for two HOV lanes as well as a light rail line.[60]

After the meeting, White approved an aggressive schedule. Construction would start in 1993 and the lane would open to traffic in 1995. He also decided that NYS DOT would consider three different alternatives (a six-lane highway, an eight-lane highway, and six lanes with an HOV lane) when the engineers began their design and environmental review process.[61] But first it was time for even more meetings and studies.

3

Finalizing Plans for the HOV Lane (1988–1995)

New York State Department of Transportation (NYS DOT) commissioners can provide critical leadership to move a highway project along a path from planning and engineering to actual construction. They can make decisions, obtain resources, and generate necessary support. Commissioner Franklin E. White realized that creating an effective HOV lane would require much more support than a typical project because local transportation officials, businesses, and organizations would have to provide ongoing transportation and marketing services that encouraged people to ride buses and share rides.

When White and Westchester County Executive Andrew O'Rourke agreed to set up an HOV/Transportation System Management Task Force, White told O'Rourke that its importance could not be overemphasized.[1] The participants in the task force included White, three county executives, George Case from the Federated Conservationists of Westchester County, Sy Schulman from the Westchester County Association, and Howard Permut, Metro-North Railroad's ambitious planning director.

At the task force's first meeting in May 1988, NYS DOT officials described the philosophy behind HOV and transportation system management: to move people rather than vehicles, and ensure that future regional transportation needs would be met without significant highway expansion or bridge construction. They explained that successful HOV lanes—like those in Houston, Minneapolis, and Washington, DC—required acceptance by commuters, employers, and transportation providers.[2]

Since White had promised O'Rourke that the task force would look closely at light rail and monorail options, the full body set up a working group to make a recommendation regarding the potential for a new rail system along the corridor. Figures 3.1, 3.2, and 3.3 show photographs of the three types of rail lines that would be considered.

During the course of the I-287 planning process, many people, from reporters to governors, confused the characteristics of these three different types of rail system. On commuter rail lines, heavy locomotives and train cars operate along dedicated rights-of-way. The New York metropolitan area has three of the nation's largest commuter railroads—Metro-North, New Jersey Transit, and the Long Island Rail Road. Light rail is similar to the streetcars and trolleys that ran in nearly every American city in the nineteenth century; the trains operate on two rails, either on their own right-of-way or on city streets. Light rail systems typically have lower capacity, fewer cars, and slower speeds than commuter railroads and can more easily be built along steep grades and sharp turns, such as those found along I-287. Monorail systems have relatively small and lightweight trains operating on a single elevated rail line. In the late 1980s, monorails were found in zoos and theme parks, including Disney World and Disneyland.

Different groups advocated for each of these systems along the corridor. The Empire State Passengers Association spoke out, wrote letters, and published newsletters calling for a light rail line along I-287. The association told the task force members that an HOV lane would be costly and ineffective and that the trend in the United States was toward light rail.[3] At the time, light rail was becoming increasingly popular. New Jersey was building a new light rail line along its Hudson River waterfront, while similar systems were being constructed in Los Angeles, Baltimore, St. Louis, Memphis, Denver, Dallas, and Salt Lake City. The association claimed that a light rail line could accommodate ten times as many people as an HOV lane. It was a specious claim, however, since buses on an HOV lane could carry just as many passengers as a light rail line.

Robert Weinberg, one of the task force participants, was a leading commercial real estate developer in Westchester County. In 1986, he led a construction industry–sponsored study to evaluate the feasibility of a monorail along I-287.[4] The study recommended building a monorail line that would serve key

FIGURES 3.1–3.3 (*facing page*) Fig. 3.1 (*top*): New Jersey Transit light rail. Light rail was among the options considered by the HOV/Transportation System Management Task Force. (*Source:* Adam E. Moreira, http://en.wikipedia.org/wiki/File:New_Jersey_Transit_Newark_Light_Rail_Kinkisharyo0104.jpg. Creative commons license). Fig. 3.2 (*middle*): Seattle monorail train. A monorail system also was considered by the HOV/Transportation System Management Task Force. (*Source:* Klaus with K, http://commons.wikimedia.org/wiki/File:Seattle_monorail01_2008-02-25.jpg. Creative commons license). Fig. 3.3 (*bottom*): Metro-North commuter train. Commuter rail was another of the options considered by the HOV/Transportation System Management Task Force. (*Source:* Adam E. Moreira, https://en.wikipedia.org/wiki/File:Sampler_of_Metro-North_services.jpg. Creative commons license).

104A
NJ TRANSIT
104A

208
208

destinations along the corridor, including Weinberg's own corporate office parks. Weinberg's business success, engineering background, and political connections added credibility to the monorail concept.

The task force's working group began its work by looking at the physical feasibility of adding a full-fledged commuter rail line over the Tappan Zee Bridge. The Thruway Authority determined that the bridge's structure was not strong enough to support commuter rail, but it could accommodate light rail or a monorail.[5] However, the bridge's new seventh lane was only wide enough to accommodate one set of tracks; building an additional track would require taking away an existing traffic lane.

The working group determined that a rail system, although physically feasible, was not practical. They realized that a successful rail line required dense urban neighborhoods or an environment in which many people were headed to one central location.

Before I-287 was built, the bulk of the corridor's population and commercial development were clustered around small, walkable communities near rail lines and the Hudson River, but that was no longer the case. For the most part, the corridor now had dispersed suburban destinations, including shopping centers and office parks, designed for automobiles. Suburban homeowners could drive directly from their driveways to parking lots next to their office buildings.

Once someone purchased a car, trips were relatively fast, inexpensive, and convenient. If a new rail line were to be built along I-287, most of its users would find their trips long and tedious. They would have to drive to a station, park the car, wait for a train, take a lengthy ride with multiple stops along the way, arrive at the station closest to their office, and then wait for a shuttle bus to take them to their final destination. For most commuters, a train would take longer and be less convenient than driving, although it would be a wonderful option to have. A shrewd commuter would continue driving while encouraging his or her neighbors to reduce traffic congestion and air pollution by taking public transportation.

The members of the working group considered routing a light rail line along Westchester's commercial streets so that it could reach downtown White Plains, but they ultimately concluded that it would be too costly, disruptive, and detrimental to local vehicle traffic.[6] The working group estimated that a rail line along the I-287 corridor would cost about $500 million to build and then would need ongoing subsidies. No public transit system in the nation since the 1960s had been able to cover all its operating costs from fare revenue. In their view, a light rail line could only be feasible if someday the trends of suburbanization were reversed and Westchester and Rockland allowed developers to build higher-density homes and offices near rail stations.[7] That is precisely what would happen along some of the light rail lines that were built in other parts of the nation during this period, but the suburban communities along I-287 corridor did not want dense urban-style development.

The working group opted instead for the HOV alternative, which would be more effective and cost much less to build and operate. Most important, it would attract more riders. Carpoolers could meet at park-and-ride lots and go directly to their office buildings. Buses could travel along the HOV lane and then go onto local streets to access the final destinations. The HOV lane would match NYS DOT's plan for the Tappan Zee Bridge, as it would be a single reversible lane—eastbound in the morning and westbound in the evening.

When the full task force met to review the working group's analysis, it agreed that a rail line should not be pursued; although it believed that rail service could one day supplement an HOV lane if and when the HOV lane approached its capacity.[8] As the monorail's biggest supporter, Weinberg decided to get behind the HOV lane concept because the median dedicated to it could one day be used for rail.

Before the task force had begun its work, Westchester County Executive O'Rourke had requested that the rail concept be studied in "sufficient detail to put it to rest, once and for all." In April 1989, his commissioner of planning was satisfied that the task force had done just that.[9] However, the idea of a rail line was not dead. It would come back again . . . and again.

Consensus Proves Elusive

Since his appointment as NYS DOT commissioner, White had made the HOV lane a priority, and he had promoted the concept when he traveled around the state.[10] Although he succeeded in getting the initiative moving, he never generated strong support for his efforts, inside or outside the task force. For example, Richard Newhouse of the Automobile Club of New York had agreed to participate on the task force, but he thought HOV lanes had only limited success in other parts of the country. Orange County's representative wanted to expand the road system, and Metro-North Railroad's Howard Permut wanted to build a commuter railroad crossing over the Hudson River.[11]

White and O'Rourke attended the fourteenth meeting of the HOV/Transportation System Management Task Force in July 1989.[12] White expected that the task force would issue a final report and "everybody would be happy."[13] He reminded the participants and the public that this was a historic effort and explained that solving the corridor's transportation problem was a big challenge given the limited amount of land available along the right-of-way and local opposition to highway projects. The emphasis, he said, must be on moving people and not automobiles.[14]

However, the task force had not developed the consensus that White had expected. O'Rourke hadn't bought into the HOV lane concept, and neither had Permut or Newhouse. Despite White's personal intervention and all the meetings of the task force and its five different working groups, positions had not changed very much. At the July 1989 meeting, O'Rourke politely thanked the commissioner and reminded him that he still wanted an HOV lane in each direction, not a reversible one.

The problem of obtaining support from the Thruway Authority continued to haunt the project. High-level NYS DOT officials believed that the Thruway Authority supported their approach to HOV lanes.[15] In fact, NYS DOT's statewide planning director, Lou Rossi, was worried that the Thruway Authority might designate an HOV lane on the bridge too soon. Rossi wanted to make sure the state had time to promote the HOV lane properly; otherwise, it might look empty, even if it was carrying more people than the adjacent traffic lanes.[16] Rossi didn't have to worry about the Authority's moving too fast, though.

NYS DOT planners working directly with the Thruway Authority understood that authority officials were continuing to view HOV lanes and transportation system management actions with extreme doubt. NYS DOT traffic engineer Rich Peters realized that the Thruway Authority was more interested in protecting its revenue stream than in promoting ride sharing. Likewise, Wayne Ugolik, an NYS DOT planner, told Rossi that the Thruway Authority was not supporting NYS DOT's ideas on ride sharing, express buses, HOV toll booths, and higher tolls during the peak periods, and he presciently warned that if the Thruway Authority did not provide an HOV lane on the bridge, a second bridge would be seen as necessary by the turn of the century.[17]

Bridge tolls were another source of tension within the task force. Ugolik attended subcommittee meetings at which Thruway Authority representatives said they did not want to see any changes that would reduce demand and lower their toll revenues. The authority staff told him they wanted to generate as much revenue as possible since their primary responsibility was to their bondholders. Despite Ugolik's reports to Rossi, the higher-ups at NYS DOT did not recognize the authority's misgivings: they attended the larger task force meetings where the press was in attendance and Thruway officials were more diplomatic.[18]

NYS DOT officials wanted to raise tolls for single-occupant vehicles, but the Automobile Club of New York representative opposed the idea. Rossi told the other task force members that he was not asking them to support a toll increase but rather a reduction in the commuter discount for those driving across the bridge alone. He had a tough time just getting the task force to agree that there should be a study of toll schedules. Likewise, a suggestion by the Westchester County Association's Sy Schulman to institute a one-cent-per-gallon gas tax to promote ride sharing and other measures in the county did not go anywhere.[19]

Schulman was becoming increasingly concerned about the future of Westchester County, recognizing that its sense of optimism had faded. Where there was once plenty of roadway capacity, now there were traffic jams. He bemoaned the fact that Westchester's communities would not allow the construction of enough affordable housing to accommodate all the people working in the county. He thought Westchester residents were once against development because it would raise taxes and crowd schools, but now, they were "just against change, development, growth, expansion, whatever it is called."[20]

Commissioner White did have support for the HOV lane from one important constituency: Westchester's environmental organizations. George Case, the president of the Federated Conservationists of Westchester County, a coalition of environmental groups, opposed a new crossing and any further widening of I-287 except for the reversible HOV lane.[21]

Final Report of the Task Force

The task force's final report, issued in September 1989, identified numerous short- and long-term initiatives, along with the dates they would be implemented and the agency responsible for them. The task force's recommendations included new express bus routes, marketing initiatives, ride-sharing programs, park-and-ride lots, TV cameras to monitor traffic, communication systems for response vehicles, and the formation of a new countywide organization to help implement transportation demand strategies.[22] Some of the short-term initiatives were either already on their way to implementation or would be implemented in the near future, including marketing efforts, new bus services, and park-and-ride lots.

The task force also approved a resolution indicating that success would be dependent upon creating a thirteen-mile HOV lane from the Thruway in Rockland County, over the Tappan Zee Bridge, and onto the Cross Westchester Expressway.[23] The task force members expected the Thruway to convert the bridge's seventh lane into an HOV lane by 1994.[24]

Peters felt that the task force had "papered over their differences" and only "agreed for show." He said "they sort of developed a consensus, but it didn't last long." He wrote to his supervisors that the task force members had seen all the resolutions of the working groups, but had never actually approved the task force's final report, let alone seen it before NYS DOT prepared and distributed it. His boss, Al Bauman, wrote back to him, "You raised an interesting point."[25]

Most of the NYS DOT officials in Poughkeepsie were traffic engineers who liked to solve traffic problems. They did not want to pursue a project that was too expensive, lacked political support, or generated widespread community opposition. They understood that the era of building new highways was over and that environmental considerations were now important. They were trying really hard to balance various interests, but that was not so obvious.

The new suburban residents of Rockland County stuck in I-287 traffic missed the convenience of the extensive public transportation network they had left behind in New York City, but it was prohibitively expensive to build and operate an extensive system of rail services to single-family homes and to office parks surrounded by seas of parking spaces. Moreover, a rail line would only attract a very small number of commuters. That did not stop rail advocates and the press from promoting it and raising the public's expectations about its feasibility.

The editorial board of Rockland County's *Journal News*, one of several daily newspapers published at the time by the Gannett Company in Rockland and Westchester Counties, did not think any of the task force's recommendations would help very much. In April 1989, the newspaper argued that the task force needed to be more forward thinking, and that it should not have ruled out the monorail and light rail ideas. In an editorial titled "Think Monorail," the newspaper wrote, "The important thing is to think in terms of new technology." A few months later, it advocated for both a light rail line and a monorail line, as well as for a new rail line to Manhattan.[26] The *Journal News* also said that a new highway bridge should be built, but not between Rockland and Westchester Counties because that would damage Rockland's waterfront. To solve the congestion problem, it suggested that the Tappan Zee Bridge, which had been designed to transport goods across the state, might have to be reserved for commuters only.

The editorial writers may have reflected community sentiment, but they exposed a willful ignorance and selfishness about solving the congestion problem. The newspaper wanted other communities to bear the burden of more vehicles while Rockland residents would get costly new rail lines that would attract relatively few users. Their position also raised the level of skepticism about the state's efforts and the feasibility of alternatives.

The editorial board seemed to have succumbed to some sort of monorail spell, similar to the one that later befell the residents of Springfield in a 1993 episode of the animated series *The Simpsons*.[27] In the key scene, a smooth-talking traveling salesman leads residents in a song-and-dance routine that seduces them into supporting a harebrained scheme to build a monorail system rather than repair their crumbling Main Street. "It's Springfield's only choice," sings the salesman, exhorting citizens to throw up their hands and raise their voices for the monorail.

The *Journal News* editorial writers claimed that the "transportation planners have to get away from road- and bridge-building. They may have been the solution in the 1950s–70s, but any more new roads will simply bring more cars and drivers while destroying more land."[28] Unfortunately, the newspaper did not understand that the state's transportation department had evolved. In a dramatic departure from its traditional focus on continually expanding highway capacity, NYS DOT's officials now were trying to accommodate growth in a way that was sensitive to the environment, and they had begun to acknowledge that the metropolitan area could not build its way out of congestion. They now understood that highway projects had social and economic impacts, that new lanes built to accommodate peak period travel would go unused at other times, and that drivers of single-occupant vehicles were demanding more space than they should be given.[29]

In 1987, Alan Bloom, a NYS DOT planning official, told the White Plains traffic commissioner that a six-lane expressway with an HOV lane was better than one with eight lanes because "mass transportation would be less attractive if the general traffic lanes operated at an acceptable level of service."[30] Bloom's comment indicated that highway congestion, NYS DOT's longtime arch-enemy, was now a good thing, because it would promote the use of carpools and public transportation.

The vehicles in the proposed HOV lane would be able to travel at fifty-five miles per hour, saving commuters fifteen to twenty minutes in travel time. The HOV lane was expected to carry 2,500 to 6,000 people per hour, while a general-purpose lane filled with single-occupant vehicles could accommodate only about 1,900 people per hour. It was even possible that one day, the HOV lane could carry tens of thousands of people per hour in buses.

Although NYS DOT had evolved, its partner, the New York State Thruway Authority, had not. The authority officials continued to resist the HOV lane, and the authority would find itself either unwilling or unable to adjust its tolls to reduce congestion.

Congestion Solutions: From Concept to Design

Once the NYS DOT decided to build an HOV lane, responsibility for the project was passed from its planners to William Kikillus, the regional design engineer at NYS DOT's Poughkeepsie office. His preliminary design effort would include surveying, mapping, soil testing, determining drainage needs and utility requirements, conducting traffic counts and traffic analyses, and preparing drawings. However, Kikillus knew that the biggest challenge for the HOV lane was political rather than engineering. Over the past twenty years, there had been a continuing decline in both transit use and carpooling. Some HOV lanes, including ones in Orlando and Los Angles, had been deemed failures.

In January 1988, Kikillus told NYS DOT's planning director, Lou Rossi, that the I-287 lane "was innovative, perhaps even daring" and that the project faced "high rewards, but also involved high risks." Kikillus realized that even the best-designed HOV lane would have limited success unless it was accompanied with incentives for commuters to share rides.[31] NYS DOT had already undertaken the construction of five hundred park-and-ride lot spaces in Rockland County, but Kikillus wanted the state to build even more.

Since NYS DOT planned to use federal funds, it was not allowed to acquire any property or make a final decision on the project until it conducted an environmental review and completed an environmental impact statement (EIS) for approval by the Federal Highway Administration. The EIS would assess the project's impact on noise, air, water, aesthetics, cultural resources, and other factors. The design and environmental review services were expected to cost over $5 million.[32]

Kikillus recognized that the project needed the support of the NYS Thruway Authority. If the authority did not designate a lane on its portion of I-287 for HOV use, an estimated 35 percent fewer vehicles would use the HOV lane on the Cross Westchester Expressway.[33] Kikillus also believed the Thruway Authority needed to offer a larger discount to carpoolers. At the time, the bridge toll was $1.50, regular commuters paid seventy-five cents, and carpoolers paid fifty cents. Since the population along the corridor was relatively affluent, it seemed unlikely that saving a few cents would motivate many of them to start carpooling.

Upon reviewing the old planning documents relating to the I-287 HOV lane, Kikillus was struck by one thing in particular. The interim report of the Tappan Zee Corridor study had indicated that the HOV lane would only provide an additional 2 percent reduction in single-occupant vehicles on top of the 18 percent from all the other transportation demand management measures. He wondered how that could that be right, given the anticipated expense and effort of the initiative.[34] Since Kikillus had not been part of the long planning exercise, he did not realize that the rationale behind NYS DOT's decision to spend hundreds of millions of dollars to build an HOV lane had relied on hope and intuition as much as technical know-how.

The complicated relationship between NYS DOT and the Thruway got even more so in the 1990s. By that time the state was facing huge budget deficits along with deteriorating highways and bridges. In a move designed to raise state funds and reduce the state's ongoing maintenance costs, in 1990 Governor Mario Cuomo's administration sold the Cross Westchester Expressway to the Thruway Authority for $30 million. The Thruway Authority would now be responsible for the expressway's day-to-day maintenance tasks such as filling potholes, clearing snow, and maintaining signs. NYS DOT, however, would continue to be responsible for the rehabilitation and reconstruction of the roadway. The Thruway Authority's contention that it could do a better job than NYS DOT maintaining the expressway added to the strain between the Thruway Authority and NYS DOT.[35]

NYS DOT even became reliant upon the Thruway Authority for financing when the state authorized the Thruway Authority to issue billions of dollars in bonds to finance projects on NYS DOT's roads. This fiscal gimmick allowed state officials to undertake transportation improvements without seeking approval from the voters to raise the state's debt limit. These bonds would not be paid back with tolls, but rather with state fuel taxes.[36]

The State of New York also obtained the funding it needed for the HOV lane on the Cross Westchester Expressway. Its congressional delegation earmarked $200 million as part of the $155 billion Intermodal Surface Transportation Efficiency Act of 1991.

A Slow and Methodical Review

Between 1989 and 1995, NYS DOT implemented some of the task force's short-term recommendations, including new bus lines and programs to promote carpooling and transit use in Westchester.[37] To please NYS DOT, the Thruway Authority opened an HOV-only toll booth on the Tappan Zee Bridge.[38] During Mario Cuomo's administration, it also worked with NYS DOT to prepare a proposal to obtain federal funding for a congestion-pricing program, but according to a former Thruway Authority senior executive that proposal is still "somewhere in the bowels of the Thruway Authority" because the governor's office shot down the idea.[39]

Planning for the HOV lane continued. NYS DOT's regional office completed its preliminary design and environmental review of the Cross Westchester Expressway HOV lane in 1995, a process that was much more complex than originally anticipated. First, the department needed to develop new guidelines for designing and evaluating HOV lanes. Then it focused on redesigning the highway and interchanges in a way that would minimize the impacts on the commercial and residential properties close to the expressway.

The state's engineers also had to address the concerns of federal officials who wanted further study of the ten-lane alternative.[40] At the time the Federal Highway Administration officials were skeptical about the HOV lane because they had never seen one that ran from one suburb to another; they were only familiar with those that ran from suburban areas into cities.[41]

NYS DOT's Peters felt that the design and environmental review took much longer than expected because of a "lack of will."[42] He felt that no one outside NYS DOT's regional office actively pushed the project forward. Others in favor of an HOV lane were only offering "grudging support." While the department's less controversial projects seemed to move along at a faster rate, the planners and engineers working on the HOV lane always seemed to have one more thing they needed to evaluate.

The project's complexity slowed its progress. Since NYS DOT's engineers wanted to minimize costs and property takings, they designed lanes to be less than the interstate standard of twelve feet. Getting approval for a nonstandard width required more studies and numerous discussions between the regional office in Poughkeepsie, the commissioner's office in Albany, and federal highway officials.

Relatively straightforward changes, such as changing the median from low metal guardrails to solid concrete barriers, changed the sight distance for drivers. That in turn required elements of the expressway to be redesigned. Peters noted that there were "so many little details that engineers spent a lot of time studying, like the width of a shoulder and the slope of a road, that you can get lost in the process."[43]

NYS DOT ruled out adding a second HOV lane, an idea that had been promoted by County Executive O'Rourke and Assemblyman Pataki, since that would have required the state to take fifty more homes and businesses along the right-of-way and cost an additional $130 million. The state's transportation planners determined that two-way HOV lanes were not needed since most vehicles were traveling east in the morning and west in the evening.[44]

NYS DOT had hoped to complete construction in 1995, but the engineers and planners were falling further behind schedule, and inflation kept increasing the project's cost. In the late 1980s, the estimated price tag for the HOV lane was about $200 million. Peters knew the estimates were likely to increase as the engineers considered all the details associated with acquiring property, widening the expressway, reconstructing overpasses, and building the lane. He had asked his bosses, "At what point does this project become too expensive, in terms of both costs and impacts?"[45] The answer to that question, he would learn, is that once the commissioner decided on the HOV lane, it would become embarrassing for the department to change course.

When Commissioner White gave his support to the HOV lane in 1987, he thought it would cost less than an eight-lane highway. If White had been presented with more accurate cost estimates at that time, he might have made a different decision. By 1995, as shown in table 3.1, the HOV lane was expected to cost about $100 million more than the eight-lane alternative.

Even if NYS DOT wanted to change the project, it seemed to be too late. The EIS process was nearly complete and the federal legislation authorizing the $200 million was designated for an HOV lane project. If the state decided to reconstruct the expressway without the HOV lane, it would not be eligible for those funds.

In May 1995, when NYS DOT officials issued their draft EIS, they still claimed that the HOV alternative was the only solution that could accommodate growth and development in the corridor.[46] It provided the best combination of transportation and environmental benefits because it would improve

Table 3.1
Cost Estimates of Cross Westchester Expressway Options

Alternative	1988 cost estimate	1995 cost estimate
Six-lane upgraded highway	$72 million	$140 million
Six lanes with HOV lane	$208 million	$365 million (not including Thruway costs)
Eight lanes	$234 million	$267 million
Ten lanes	$287 million	Not estimated

SOURCES: Costs estimates from 1988 are found in NYS DOT, I-287/Cross Westchester Expressway Development Study." The 1995 cost estimates are in Federal Highway Administration and NYS DOT, *I-287/Cross Westchester Expressway: Final Design Report/Final EIS*, V-8.

safety, mobility, efficiency, and air quality without increasing the number of lanes for single-occupant vehicles. Their traffic models showed that during peak periods, vehicles in the HOV lane would travel at fifty-five miles per hour while those in the general lanes would travel at thirty.

NYS DOT claimed that the HOV project had "strong support in Rockland and Westchester" and that it significantly bore the "stamp of public input and approval."[47] In reality, those claims were wishful thinking. Aside from the construction firms that supported every transportation project proposed by the state, the project had few enthusiastic supporters.

Ultimately, given the ongoing public debate and the complexity of the design, the I-287/Tappan Zee Bridge HOV lane would not be the state's first to open. That distinction would go to the HOV lane on the Long Island Expressway, which became operational in 1995. Long Island's HOV lanes were easier to build because the state owned more right-of-way, the terrain was flatter, and the interchanges were spaced further apart. The HOV lane spanning Rockland and Westchester was a different kettle of fish altogether, and it would face unexpected opposition in the days ahead.

4

Killing the HOV Lane (1994–1997)

When New York State Department of Transportation (NYS DOT) engineer Rich Peters first started planning improvements for I-287 in 1980, HOV lanes were considered an innovative solution to traffic congestion. The Hudson Valley's environmentalists were among the few stakeholders enthusiastic about NYS DOT's efforts to improve traffic conditions without building new highways or widening existing roads for single-occupant vehicles. However, their support would prove to be fleeting when a new generation of activists emerged to oppose any project that widened highways, even if the new lanes were dedicated to carpoolers and buses.

The Tri-State Transportation Campaign

George Case represented the environmental community on the HOV Task Force study during its fourteen months of deliberations in 1988 and 1989. Case was president of the Federated Conservationists of Westchester County, an organization that provided a unified voice for more than three dozen local Audubon Societies, garden clubs, conservancies, nature centers, and advocacy groups. Case agreed to serve on the task force only after he was reassured that it would not consider a new bridge.[1]

Case supported the task force's recommendations that a monorail system was not practical, and that building a light rail line on local roads would be disruptive to communities. However, he considered an HOV lane to be an effective way to solve the vexing transportation problem. To Case, the concept was environmentally progressive, effective, and relatively inexpensive. The state would need to add only one lane in the existing right-of-way and it would require few property takings. In 1989, he explained to his members the benefits of an HOV lane—it would allow buses and carpoolers to pay lower tolls, go through their own tollbooths, and travel in their own lanes across the Tappan Zee Bridge and in Westchester.[2]

In the early 1990s, environmentalists in New York City began to see HOV lanes differently. Jim Tripp, an attorney at the Environmental Defense Fund, was concerned that the three states in the New York metropolitan area were spending too much money on expanding highway capacity. He thought more funds should be allocated toward improving public transportation while simply keeping the existing highway system in a state of good repair. Tripp wanted to break what he called the "never-ending cycle" of building more highway capacity that would lead to more cars, more congestion, and the need to expand highways ever further.[3]

Tripp remembered a lesson he learned in the 1970s from William Vickrey, a Columbia University professor who was later awarded the Nobel Prize in economics. Vickery explained to Tripp that economic incentives and disincentives could avoid the need to build new power plants. Instead of increasing capacity, the utility companies could charge customers higher prices during peak periods and lower prices during off-peak periods. Tripp was amazed to learn from Vickery that the concept could also be used for transportation, a concept that he would remember years later.[4]

In 1990, the Environmental Defense Fund could have gone to court to try to stop some of the highway expansion projects in the New York metropolitan area, but the politically savvy Tripp believed that litigation would only delay the projects and serve to alienate both the public and elected officials. Instead, he had another idea.

Tripp started bringing together well-connected advocates from the region's leading environmental and public transportation advocacy groups (such as the Straphangers Campaign and Project for Public Spaces) to meet at the Environmental Defense Fund's office every month. They discussed how they could work together to promote environmentally friendly transportation policies. After Congress passed the Intermodal Surface Transportation Efficiency Act of 1991, which tied federal transportation funding more closely to environmental requirements and gave states the ability to use federal highway funds for transit improvements, Tripp began formalizing the relationships he had developed.

Under Tripp's leadership, the participants started a new nonprofit organization, the Tri-State Transportation Campaign. Rather than creating yet another environmental advocacy organization, Tripp wanted Tri-State to be seen as an independent, business-friendly organization that focused on improving the region's transportation system.[5] After receiving funding from three different foundations to set up and run the organization, Tri-State hired Janine Bauer, an attorney with a psychology degree, as its first executive director.

To Tri-State's founders, highway expansion was not a solution to transportation problems; rather, it was the problem that was exacerbating air pollution and congestion.[6] As they began compiling a list of the region's planned highway projects, they began to think that the transportation departments were adding new highway lanes "under the subterfuge of High-Occupancy Vehicle lanes."[7] Tri-State would have supported transportation projects that converted general use highway lanes to HOV lanes, but it was opposed to HOV lanes that actually widened highways.

At the organization's first annual retreat, Tri-State's founders spent two and a half days discussing and arguing about the various initiatives that they should undertake, and how they should allocate their grant funds. Since they did not have the resources to fight every highway capacity expansion project in the metropolitan area, they decided to focus on issues that would achieve the greatest impact. The high-profile I-287 HOV lane would be their number-one target.[8]

Tri-State quickly established itself as an influential player in the transportation policy arena. It started a weekly newsletter called *Mobilizing the Region* that was sent to transportation professionals and media in the New York metropolitan area. Only four pages long, it could be read in about five minutes. It filled a void: no other media outlet was reporting on the region's transportation projects in the same breadth and depth.[9] Tripp found the newsletter's impact was phenomenal; it seemed that virtually everyone in government, academia, nonprofits, and business who was interested in transportation issues was reading it. He said, "It created a sense of transportation interconnectedness that had not really existed in the public consciousness."[10]

The first issue of *Mobilizing the Region* in September 1994 described activities Tri-State was taking to orchestrate opposition to the I-287 HOV lane. The next month, Tri-State started faxing the newsletter to more than three hundred transportation professionals, elected officials, citizen activists, business leaders, and reporters.[11] By the end of 1995, Tri-State was faxing *Mobilizing the Region* every week to more than fourteen hundred people.[12]

Although most of the people who had been planning transportation improvements along I-287 had seen more roads as a way to promote economic growth, Tri-State's founders thought more roads would have the opposite effect in the long run, fueling an ominous sprawl of automobile-dependent, low-density development.[13] Tri-State supported exactly the same type of alternatives recommended by the I-287 transportation studies in the 1980s—charge higher tolls during peak periods, increase transit services, offer flexible work schedules, and eliminate free parking at office parks.[14]

As Tri-State's executive director, Bauer became the leader of the fight to stop the I-287 HOV lane. In 1994, she asked the state to enter into a collaborative planning process to develop solutions to the transportation problems along the Cross Westchester corridor. But after more than a dozen years of studying the corridor, the state had no interest in starting over with its planning process.

Businesses, residents, and local officials were concerned about the project's construction impacts, such as traffic diversions, noise, and property takings. Tri-State publicized these impacts and intensified people's fears. The lessons Bauer learned in law school came in handy. She used NYS DOT's own facts to expose the project's weaknesses. "That's what lawyers do in court," she said.[15]

Under Bauer's leadership, Tri-State emphasized different aspects of the HOV lane to various Westchester County audiences. Bauer explained to Westchester's legislators that factories were no longer the main source of pollution, that most of the region's air-quality problems were now emanating from cars and trucks. She showed them graphics of "unsightly fly-overs and ramps" that NYS DOT needed to build to accommodate high-occupancy vehicles.[16] The business community started to worry after Tri-State explained to them how the multiyear construction process would hurt the county's economic well-being.[17]

Tri-State convinced the communities along the expressway that "they were growing screwed."[18] Westchester's communities started to see the expressway's expansion as a project that would mostly benefit people who lived west of the Hudson River at the expense of Westchester residents. Tri-State also emphasized the high cost of building the HOV lane. Although NYS DOT estimated the HOV project would cost approximately $365 million, Tri-State pointed out that it would probably be close to $500 million because actual construction costs were typically underestimated, and those estimates did not include expenses the Thruway Authority would have to incur to build an HOV lane on its portion of I-287.[19]

Bauer was able to outmaneuver NYS DOT thanks to the complementary skills of three energetic and talented people.

Michael Replogle was brought in from Washington, DC, by the Environmental Defense Fund. He was a self-styled hired gun, empowered to find holes and exploit weaknesses in the state's HOV analysis. A civil engineer and sociologist by training, Replogle understood the complex traffic-modeling tools that the state used to justify the HOV lane. He considered NYS DOT's methodology to be primitive because it did not consider how building the HOV lane could lead to more traffic, growth, and sprawl.[20] Replogle argued that adding an HOV lane would induce more people to move to the exurban and semirural areas west of the Hudson River, which would sabotage the economic redevelopments efforts of Westchester's urban and village centers.[21]

Replogle could see that Peters and NYS DOT were trying to manage traffic in an environmentally sensitive manner, but he and the Environmental Defense Fund wanted to draw a line in the sand on highway capacity expansion. Their fundamental objective was to implement a region-wide congestion-pricing program and dedicate funds from such a program to transit services. They wanted the three states to set tolls throughout the New York metropolitan area that would be high enough to keep traffic moving smoothly all day. That would improve air quality and generate funds for a high-quality public transportation system. Replogle brought in a representative from Trondheim, Norway, who explained how his city was installing tollbooths throughout its road network in order to reduce congestion and finance new transportation projects.[22]

The second key player on Bauer's team was Jeff Zupan, recruited by Tripp because he was a well-respected transportation professional from the highly regarded Regional Plan Association. The association, led by leading members of the business, philanthropic, civic, and planning communities, had been proposing plans to guide the growth and development of the New York area since the 1920s. The association advocated strengthening commercial business districts, protecting the environment, and minimizing sprawl. Before joining the Regional Plan Association, Zupan had coauthored three books about transportation and headed New Jersey Transit's planning department.[23]

Zupan opposed the HOV lane because he thought it would be lightly used and that there would be an outcry to open it up to all traffic.[24] He had good reason for his concern; by 1995, HOV lanes had been converted to general-purpose lanes in Atlanta, San Francisco, San Diego, Ventura, and northern Virginia.[25] Zupan also thought the HOV project was an inefficient use of limited transportation resources, since according to NYS DOT's own analysis, most drivers still would be stuck in traffic on the Cross Westchester Expressway a few years after the HOV lane opened.[26]

Although some other Tri-State members, such as Janine Bauer and Jim Tripp, were against almost all highway capacity expansion projects in the metropolitan area, Zupan and the Regional Plan Association were not. Zupan thought certain highway improvements could effectively relieve perpetual bottlenecks and complete missing links when there was no obvious transit solution. On I-287, he saw both the need to make specific highway improvements and an opportunity to implement a successful demand management program.[27]

Zupan's own traffic analysis revealed the existing congestion problem stemmed from two distinct factors. The closely spaced interchanges on the Cross Westchester Expressway created excessive weaving over short stretches, and the obsolete design of the Thruway's intersection with the Cross Westchester Expressway was causing bottlenecks. He calculated that the state could significantly improve traffic flow without widening the road if it could take a small number of cars off each segment of the expressway. He thought the Thruway Authority should start reducing demand by changing its "perverse" tolling policy in which the authority gave discounts to regular commuters who typically traveled during peak periods. Instead of giving them a discount, he wanted those rush hour drivers to pay a premium.[28]

With Zupan and Replogle, Tri-State had the technical expertise to critique and expose the vulnerabilities of NYS DOT's analysis. But they needed someone who could be a highly effective community organizer in Westchester County. So Jim Tripp called upon another friend.

Tripp had met Maureen Morgan when she worked as a secretary at the Environmental Defense Fund. After she had been elected president of the Federated Conservationists of Westchester County, Tripp asked her if the Tri-State Transportation Campaign could make a presentation to her organization. In early 1994, Bauer and Tripp came to Morgan's second board meeting as president of the county conservation group.

Morgan turned out to be the effective local voice the New York City–based environmental and transportation advocates had been seeking. Years later, Morgan said, "I didn't know anything about transportation at that time. I also didn't know that I was in for four years of heavy combat."[29]

When George Case had been the Federated Conservationists president and served on the HOV Task Force, he supported the reversible HOV lane. Morgan would come to hold the opposite opinion. The more she learned about the lane, the more convinced she became that "it did not make much sense." She felt "carpooling was dying" because no one was working from nine to five anymore. She also thought the HOV lane would only help a handful of people while everyone else would still be stuck in gridlock.[30]

Starting in 1994, Morgan explained to numerous businesses, civic groups, and local communities why they should oppose the HOV lane project. She wrote op-ed pieces and appeared on cable TV shows. She hosted public events, forums, and conferences and met with county legislators, the county executive, and candidates running for public office.

Morgan was a tireless advocate who thought writing, organizing, debating, and appearing in public was "a lot of fun. I had a good time and I had nothing to lose." When the NYS DOT regional director responded in the newspaper to one of her op-eds, it emboldened her further. She remembers, "I liked it when I was attacked. It gave me an opportunity to counterattack and it gave me a thicker skin."[31]

Having previously worked as a choirmaster and music director, Morgan found that she was putting many of the same skills to work as an advocate. She said, "You have to get everybody to sing on the same page. I had worked with all kinds of different people—from certified crazies to kids to juvenile delinquents. I always made people feel comfortable because I don't come off like someone above others."[32]

After she brought the HOV lane issue to the Westchester County Board of Legislators, that body became an important supporter of Tri-State's efforts. One legislator taught Morgan how to write a resolution, a skill that helped her get resolutions passed against the HOV lane in communities all across the corridor.[33]

In November 1995, the seventeen-member county legislature unanimously adopted a resolution using arguments crafted by Tri-State.[34] The resolution stated that the HOV lane would bring more concrete, pollution, and noise, it would not relieve congestion, and it would severely disrupt normal traffic flow for six or seven years. The legislature said that money set aside for the HOV lane should be spent on public transit alternatives for Westchester's residents. Adding lanes, they argued, would exacerbate inefficient land use and encourage further automobile dependence.

Morgan may have been "public enemy number one" to NYS DOT officials, but Peters met with her at NYS DOT's office in Poughkeepsie. It was painful for both of them.[35] Peters was a true believer—a well-intentioned expert who had put countless hours into planning the HOV lane. He had even used his own vacation time to attend HOV conferences. When they met, Morgan looked at the analysis and the drawings that Peters had prepared and found them very moving, like a piece of artwork. She knew that Peters had worked on the project for more than a decade and was really passionate and committed to it. Still, she found the scheme totally impractical.[36] Peters meanwhile felt that Morgan and the general public never really understood the project's benefits.

Peters's honesty was hurting his project. Some proponents of transportation initiatives will manipulate information to help advance their goals. It would have been relatively easy for Peters to change a few numbers to put the HOV lane in a better light. For example, he could have estimated that more people would start carpooling and riding the bus after the HOV lane opened. Then he could have shown how the HOV lane would significantly reduce congestion for all drivers. Instead, Peters gave his opponents ammunition with which to attack him.

Opposition Mounts

The Tri-State Transportation Campaign started to remove the key pillars from the project's shaky foundation of support. It influenced both the public reaction to the HOV lane and the media's coverage of the project's flaws. The HOV lane's opponents also skillfully raised the public's expectations about how other ideas could solve the congestion problem. After the state released its draft environmental impact statement (EIS) in May 1995, opposition to the project grew more intense.

In a September 1995 editorial, Gannett's *Citizen Register* newspaper called the HOV lane unworkable.[37] It decried "unsightly high ramps and interchange systems" and claimed that it would not have enough entrances and exits to serve commuters. In December 1995, the day before a Westchester public hearing on the project, an editorial in Gannett's *Reporter Dispatch* urged residents to tell NYS DOT that the project would be disruptive and would shift traffic congestion to local streets, and it urged state highway planners to find new routes to link New England with the rest of the country.[38] The editorial writers told the state that it should go back to the drawing boards and build a transit project that would have far-reaching impacts.

At public hearings in December 1995, many of the attendees were skeptical about the effectiveness of an HOV lane. For example, Westchester County Planning Board chairman William Cassella said a creative approach to improving transportation would uncover solutions that were safer, less costly, more environmentally friendly, and more effective.[39] Representatives of communities along the right-of-way opposed the project because the state would take private property, residents would be subjected to a noisier roadway, and construction would divert traffic to local roads for several years.

An opponent named Frank Ronnenberg especially infuriated Peters. NYS DOT had helped Westchester's business leaders establish a program to promote carpooling and transit for employers in the county.[40] Even though NYS DOT was paying Ronnenberg's salary to head that program,[41] Ronnenberg testified that the state should use the hundreds of millions of dollars allocated for the HOV lane for better transportation projects. He especially did not like the reversible aspect of the HOV lanes and said, "There is more planning to be done."[42]

All of the state legislators who testified at the hearings criticized the project. An assemblywoman said the HOV was not a solution to a long-range problem and would just attract more vehicles and trucks. She preferred the six-lane alternative with expanded bus services.[43] A second Westchester assemblywoman had been a supporter of the HOV lane but had changed her mind. She argued that most residents using the HOV lane would not be from Westchester, while Westchester residents would bear the burden of the construction.[44] A third assemblywoman said that losing $200 million in federal funds if the lane was not built was a myth that had been told over and over again. She said the state could indeed use the $200 million for something better, like a light rail line.[45]

Nevertheless, the HOV lane still had a few fans left. Bob Weinberg, developer and monorail advocate, deplored the pretense of those who suggested doing nothing would magically create the infrastructure needed to provide public transportation. He argued that if county residents wanted to accommodate more people via carpools or rail, the state needed to widen the existing right-of-way.[46] Meanwhile, Ross Pepe, who in 1978 started the New York Construction Industry Council of Westchester and Hudson Valley, a trade organization representing hundreds of construction firms and tens of thousands of workers, said the corridor would be "traffic free" with the HOV lane.[47]

Peters believed that both NYS DOT and the Tri-State advocates simply wanted safe, economical, and efficient transportation and that they were "just disagreeing sometimes on how to get there."[48] He realized, by the time it was too late, that NYS DOT had been outmaneuvered by the environmentalists. When Peters visited legislators' offices to explain the benefits of the HOV lane, he discovered that Tri-State representatives had already been there. Although Peters was well regarded by state officials and the Tri-State advocates for his diligence and technical skills, he did not have the political savvy or marketing know-how to counter their arguments effectively.

Peters was getting frustrated. He told Replogle after a meeting, "You're going back to DC when it's over and I'm stuck with this problem."[49] He lost his temper at another meeting when Morgan said she wanted more bus services but no new lanes. "Do you think buses fly?" he snapped.[50]

Year after year, study after study, Peters had shown that certain options were just not feasible. He had tried to explain to whoever would listen that traffic on the expressway could not be improved without taking some private property. He also had tried to convince public officials and civic groups that light rail was not feasible given the low demand and high cost for transit services. He had carefully explained that the Tappan Zee Bridge had room for only one light rail track, which meant trains could run only one way in the morning and evening. Since the light rail operator would not be able to bring equipment back and forth during the peak periods, it would have to purchase, maintain, and operate almost twice as many rail cars compared to a light rail system with two tracks. Peters also had explained how it would be difficult to attract passengers to a light rail system in Westchester County because people would have to cross wide roads and go through vast parking areas to get to their office buildings.

No matter what he said, Peters was unable to effectively address the environmentalists' concerns about sprawl because land use, planning, and zoning issues were under the jurisdiction of the towns and cities. He believed that the environmentalists were unfairly targeting the state's transportation projects because they could not stop what he considered to be the underlying causes of suburbanization and sprawl—racism and the poor quality of urban schools.[51]

Peters reported to Al Bauman, the director of NYS DOT's regional office for the Lower Hudson Valley. Bauman had extensive experience in constructing highways but he had neither Peters's extensive knowledge of the transportation network nor NYS DOT Commissioner White's vision for transforming the transportation system. He certainly did not have the community organizing skills of Morgan. In the I-287 HOV fight, he was in over his head.

When Bauman started as regional director in 1989, he began chairing the HOV Task Force, which he thought was unable to develop a cohesive set of recommendations because there were too many members. He remembers, "Quite frankly, the task force was doomed to failure from the very beginning because of the organizations represented. They were not interested in a multimodal solution, but only their own narrow and unrealistic solution."[52]

Bauman felt the ideas generated by the task force and later by the environmentalists were not practical. In his opinion, congestion pricing was unrealistic and unfair to people who had to get to work during the peak traffic periods. He believed that the environmentalists' ideas of promoting bicycles and rail were utopian. He said, "They forget people have to get to work" and noted, "We couldn't get support from newspapers. I don't know why. They seemed to think that carpools and other magical things that wouldn't work, would."[53]

Despite her flair for orchestrating public opinion, even Morgan found it hard to keep people interested and energized about the HOV lane, year after year. She was starting to get discouraged in her fight against NYS DOT when a state assemblywoman mentioned to her that none other than the New York State Thruway Authority also opposed the HOV lane plan. That news rejuvenated Morgan and gave her the boost that she needed to keep going.[54]

Thruway Authority officials believed that an HOV lane would harm their organization's image and its bottom line, as well as their careers. When Morgan subsequently met with them, she happily noted that they treated her with great respect. "They loved what I was doing, but they couldn't say it. I was their hero," she recalled.[55]

As NYS DOT's plans fell further under siege, the Thruway Authority's grudging acquiescence to the HOV lane slipped away. When NYS DOT officials had decided in 1987 to build a reversible HOV lane on the Cross Westchester Expressway, they expected the Thruway Authority to do the same on its portion of I-287 including the new seventh lane on the Tappan Zee Bridge. However, by 1995 the Thruway Authority not only opposed the HOV lane on the Thruway and the bridge but was now against an HOV lane on the Cross Westchester Expressway as well.

The Thruway Authority's engineers felt there was no justification for an HOV lane. After reviewing NYS DOT's analysis, one engineer said, "The riders just weren't there." Another engineer remembers, "If DOT built the HOV, it would have caused the mother of all traffic jams in Rockland County and the Thruway Authority would have gotten the blame." Likewise, the authority's executive director, Stephen Morgan, wanted "to give the vast majority of people the right to use the facility." He did not think that there was "enough market share to set aside a lane" and believed that enforcement would have been difficult. Thruway Authority officials were skeptical that many people would carpool, especially after the state had not been able to build as many park-and-ride lots as it wanted due to community opposition.[56]

Thruway Authority officials did not want to be embarrassed by building an HOV lane that would carry few vehicles. They felt that their customers deserved a higher level of service than drivers on NYS DOT's roads because Thruway drivers had to pay tolls. The authority's engineers expected drivers would complain about an HOV lane so much that the authority would have to convert it to general use.[57]

Financial considerations were another important factor in the authority's decisions. A lightly used HOV lane on the Thruway would bring in much less revenue than a heavily used general-purpose lane. Therefore, from the Thruway Authority staff's perspective, it was much less risky to build a new lane that could be used by all vehicles.[58]

After the December 1995 public hearings, the Thruway Authority's chief engineer, Leonard DePrima, told NYS DOT's Bauman that there was still no clear consensus among travelers and professionals on a preferred alternative, and that more consideration should be given to the eight-lane concept. He suggested that the two agencies conduct a survey of the Thruway's daily customers, asking them about their preferences and the likelihood that they would use the HOV lane.[59]

NYS DOT's desire for the Thruway Authority to raise tolls for single-occupant vehicles during peak periods on the Tappan Zee Bridge was not going anywhere either. One Thruway Authority official said, "We didn't have the stomach" to do it."[60] Although they cooperated on many initiatives, NYS DOT officials often found it frustrating to work with their Thruway Authority counterparts. One NYS DOT commissioner, John Egan, did not even like to utter the words "Thruway Authority." Instead he referred to the authority as "Brand X."[61]

After the December 1995 public hearings, the Tri-State Transportation Campaign continued to organize the opposition against the HOV lane. Morgan stepped down from her role as president of the Federated Conservationists of Westchester County, which gave her even more time to devote to mobilizing opponents.[62] The next year and a half would bring the controversy to a close.

During the long planning process, Westchester County Executive Andrew O'Rourke had often been a thorn in NYS DOT's side. At various times, O'Rourke supported an eight-lane highway, two-way HOV lanes, and a light rail line. Surprisingly, as opposition mounted from the Westchester County legislature, environmental groups, and the communities alongside the expressway, O'Rourke had become one of the last and most important supporters of the HOV project.

A January 1996, Gannett published an editorial titled "O'Rourke Heading Wrong Way on HOV" saying it was mind-boggling that the county executive would support the HOV project given its cost, physical impact, and doubtful effectiveness.[63] The newspaper suggested that the state rethink its proposal and come up with something better. O'Rourke wrote back that although he would be happy to support a better plan to ease congestion on the Cross Westchester Expressway, "unfortunately, no such plan exists."[64]

From O'Rourke's perspective, the expressway had helped fuel the county's economy, and an overcrowded highway would deter businesses from locating in Westchester. In 1996, O'Rourke did not think the state's proposal was perfect. He still preferred HOV lanes in each direction and would have supported an expansion to eight lanes, but he was concerned that if the HOV lane was rejected, the state would lose the $200 million in federal funds that were specifically earmarked by Congress for a Cross Westchester Expressway HOV lane. Without these federal funds, he feared that no improvements at all would be made.[65]

In 1988, NYS DOT Commissioner White had hoped to begin construction by 1993 and complete the project in 1995. However, by 1996, NYS DOT and its consultants still were preparing the final EIS that addressed all the concerns and questions raised by the public on the draft environmental document.

HOV Lanes Lose Some Luster

The I-287 HOV lane may have been a new and innovative concept in the New York metropolitan area when NYS DOT planners first proposed the idea in the 1980s, but by the mid-1990s, HOV lanes were losing their luster. Peters had long worried about the fatal flaw of an HOV lane—that it could look empty to drivers moving slowly in general-purpose lanes, even though it might be carrying more people.[66] That is exactly what was happening along a stretch of I-287 where the State of New Jersey had built an HOV lane. A January 1997 *New York Times* article titled "Speak Up If HOV Lanes Were Your Idea" reported that commuters were fuming "as they pile up in snarled traffic, watching the handful of certified HOV-lane users sail by like swells through a velvet rope at a nightclub."[67]

In July 1997, NYS DOT issued its final EIS as the opposition was getting even stronger. Three national environmental organizations jointly named the HOV lane one of the thirty-seven most wasteful road projects in the nation. The Gannett newspapers came out against the project again. Their editorials argued the HOV lane would encourage more traffic, worsen sprawl, hinder transit, and exacerbate auto dependence. Moreover, both men running to replace O'Rourke as county executive in 1997 announced their opposition to the project.[68]

In a final blow, a few weeks later the Tri-State Transportation Campaign launched TV spots and sent a mailing to thirteen thousand people urging them to tell Governor George Pataki to kill the project and redirect the $200 million in federal funding to alternative strategies.[69]

In the spring of 1997, the Tri-State Transportation Campaign arranged a meeting with Governor Pataki and several of his aides. Bauer, Zupan, and Morgan all met the governor in a reception room next to his office.[70] As a former legislator whose district spanned both sides of the Tappan Zee Bridge, Pataki understood the problems facing the corridor. As an assemblyman, he had preferred two HOV lanes on the expressway rather than the reversible one-way plan. As governor, he was frustrated by the Thruway Authority's inability to solve the recurring rush-hour traffic jams on the Tappan Zee Bridge.[71]

The Tri-State representatives told Pataki that Westchester's employers were not doing enough to encourage carpooling and transit options. As Zupan was explaining that taking a relatively small number of cars off the highway could reduce congestion, the governor interrupted to say, "Like when teachers are off in the summer."[72] At that moment, Zupan realized that Pataki understood Tri-State's arguments.

After Tri-State gave Pataki a list of organizations opposed to the HOV lane, the governor nodded his head but gave no commitments. He indicated, though, that he might oppose the project if the environmentalists could show they had the business community on their side.[73]

In October, Pataki met again with Tri-State and the Federated Conservationists, this time in a conference room next to the Westchester County executive's office. The governor did not bring any representatives from NYS DOT or Westchester County, which meant there was no one in the room who would defend the HOV project.

Larry Dwyer, the president of the Westchester County Association, attended the meeting. He and the governor had known each other since the 1980s when they were both elected officials in Westchester. As the head of the county's leading business association, Dwyer's opinion was important, but he had been frustrating both HOV supporters and opponents because he had not taken a public stand for or against it. He now told Pataki that the HOV lane would not serve enough people, it was impractical, and there were not enough entrances and exits for its users. When Dwyer said he was in favor of killing the HOV lane and had been against it for some time, Morgan "could hardly control a loud guffaw." She later said, "You would have thought Larry was always against it." It looked to her like "rats jumping off the sinking ship."[74]

When Pataki reminded Tri-State about the safety improvements that were part of the project, Zupan explained there were more effective ways to solve the safety problems without having to spend hundreds of millions of dollars on an HOV lane. Pataki did not think the HOV lane was a bad idea, but he did think there might be better alternatives. As an environmentalist and rail advocate, he sympathized with the transit advocates. As a former Peekskill mayor, he remembered that state highway officials sometimes did not understand issues as well as local residents.[75] After listening to Zupan's response, an exasperated Pataki shook his head and said, "I don't know if I was sold a bill of goods by DOT, but I don't think I need to stake my name on this."[76]

Pataki thought that I-287 was a critical corridor facing a "mini-crisis." He wanted the state to develop a comprehensive solution that would dramatically improve Westchester County and make it less reliant on automobiles. He thought the HOV proposal was a "creative and good idea for the moment" but that it lacked vision.[77]

Pataki Presses Restart

On October 20, 1997, Pataki announced his opposition to the HOV lane and established an I-287 Task Force to "foster a consensus on how to improve transportation, promote economic development, and protect the region's environment."[78] In effect, he restarted the I-287 planning process seventeen years after it had first begun.

Pataki said the state would try to use some of the $200 million earmarked by Congress to improve ramp bottlenecks and weaving sections, a project that was expected to cost about $150 million. He also said the solution to the corridor's congestion problem included adjusting peak-period tolls on the bridge. However, the governor ultimately would not be willing to raise peak-period tolls for cars, nor would federal transportation officials allow New York to use its federal earmark for a project that did not include an HOV lane.

Politically, Pataki made the right decision. The Tri-State Transportation Campaign, the media, and local officials had all raised the public's expectations that a better solution existed. The public perceived the reversible HOV lane as an ineffective and clumsy way to solve the corridor's congestion problem. During a construction period that would have lasted at least seven years, the HOV lane would have been seen as a disruptive nuisance. Even if an HOV lane were the best long-term transportation solution, many drivers would have considered it to be an expensive and underused boondoggle.

Tri-State's Bauer was delighted with Pataki's announcement. She said, "This represents an absolute sea change in the way these things are studied." She also stated, "Until now, traffic engineers have basically been unable to help themselves because of their schooling. They just think that to move traffic you have to lay more asphalt."[79]

Not everyone was happy with the governor's decision. A highway industry publication, *Toll Road News*, bitterly reported that this "was a major victory for anti-highway groups over state transportation planners." The editor wrote, "The task force is unlikely to achieve anything since all the options have already been exhaustively listed, examined, studied, modeled and argued over during more than ten years of efforts to achieve political support for some action." He also argued that the environmentalists "have become more rabidly anti-road and oppose anything which eases the life of the motorist, apparently convinced that misguided drivers must be punished for their selfish adherence to an evil machine."[80]

Ross Pepe, the president of the trade organization representing hundreds of construction firms, was also furious with Pataki's decision. Pepe vowed that he would not let the Tri-State Transportation Campaign beat him again.[81] Fifteen years later, they would face off against each other in an even bigger battle over the future of the I-287 corridor.

5

Permut's Rail Line and Platt's Bridge

In 1997, when Governor George Pataki canceled the HOV lane and set up the I-287 Task Force to recommend new transportation solutions, only his inner circle of advisors knew that he was considering combining plans from the Thruway Authority and Metro-North to replace the Tappan Zee Bridge *and* build a new east-west rail line along the thirty-mile I-287 corridor.

While New York State's Department of Transportation (NYS DOT) was planning its HOV lane across the Tappan Zee Bridge, two other state transportation agencies had been working on much more ambitious strategies to help solve the I-287 congestion problem. Metro-North Railroad, a subsidiary of New York's Metropolitan Transportation Authority (MTA), was planning a new rail line along the corridor while the New York State Thruway Authority hoped to completely replace the Tappan Zee Bridge with a wider structure.

A new Tappan Zee Bridge would be the grandest transportation structure built in the New York metropolitan area since the Verrazano-Narrows Bridge connected Staten Island and Brooklyn in 1964. Building a new railroad line across the Hudson Valley would be an even bolder initiative. Together, a new three-mile bridge and thirty-mile rail line would be a monumental undertaking, a once-in-a-lifetime project.

Metro-North vice president Howard Permut and Thruway executive director John Platt were the visionaries behind the two plans. Permut's notion of a new rail line across the Hudson Valley and Platt's idea to replace the bridge (with private sector participation) captured the attention of Governor Pataki, who wanted to promote economic growth, reduce automobile dependence, tie together existing rail lines, alleviate traffic congestion, foster new public-private partnerships, and improve access to Orange County's Stewart Airport.

Platt resigned from the Thruway Authority in 2003, but Permut was one of the most important players in the I-287 saga from 1989 through 2010. Not coincidentally, they both had planning backgrounds. Daniel Burnham, a visionary architect whose 1909 master plan for Chicago inspired generations of planners, once reportedly said, "Make no little plans; they have no magic to stir men's blood . . . make big plans; aim high in hope and work."[1] Permut and Platt both made big plans and had high hopes.

They also both worked at public authorities. Princeton University's Jameson Doig, the author of several books about transportation, politics, and leadership, has studied the growth and power of American public authorities. He observes that they often attract leaders who know how to set long-term goals and marshal the organization's significant resources to achieve them. These leaders often promote large-scale projects (such as new bridges and rail lines) rather than more modest and environmentally sensitive ones (such as HOV lanes).[2]

Howard Permut's Plan

Metro-North Railroad operates three north-south rail lines between Westchester County and Manhattan's Grand Central Terminal. The train service between the suburbs and the city is frequent and fast. Metro-North also provides equipment and funds for New Jersey Transit to operate service west of the Hudson River in Rockland and Orange Counties—a service that is not as convenient since passengers destined for New York City need to transfer in New Jersey for the trip across the Hudson River.

The MTA's operations are heavily subsidized by sales, mortgage, and business taxes that are paid by residents of the counties in which the MTA provides services. In the late 1980s, Rockland and Orange County officials complained they were not getting their money's worth and threatened to withdraw from the MTA service area so that its residents would no longer have to pay these taxes.[3]

Rockland County claimed that the MTA was collecting $15 million a year in taxes from its residents, but only providing $1 million worth of transportation services. After New Jersey Transit proposed eliminating Saturday train service in 1989, Orange County's transportation coordinator, Carl Daiker, said, "We run into a problem with New Jersey Transit. They can dictate what kind of service we can get. We want to find some way of getting across the river so we won't have to deal with New Jersey Transit."[4]

As head of Metro-North's planning department, Permut found it frustrating to work with New Jersey Transit since he had only limited ability to change schedules and improve service. After an MTA Board meeting at which board members from Rockland and Orange Counties were complaining about their inferior service, Permut asked one of his senior planners, Marty Huss, to identify steps they could take to improve service and bring the two counties "out of the dark ages."[5]

After evaluating Metro-North's operations, Huss realized the railroad could only provide minimal improvements given the physical and operational constraints on the lines running through New Jersey. The only way to dramatically improve service would be to build a new rail bridge or tunnel across the Hudson River. Huss said that it would be "light years ahead of anything else."[6] A new rail crossing would provide Metro-North's customers who lived west of the Hudson River and worked in New York City with shorter travel times and a "one-seat ride" because they would no longer have to transfer in New Jersey.

In 1989, the MTA Board of Directors approved Permut's request to hire planning consultants and begin a study titled "Feasibility and Benefit-Cost Study of Trans-Hudson, Cross Westchester and Stewart Airport Rail Links."[7] In the first year of the study, the planners considered thirty different alternatives. Over the next three years, they developed more refined cost and ridership estimates for six of those. Four of the six options involved new rail crossings in the Tappan Zee Bridge area. Just as Governor Dewey wanted to build the Tappan Zee Bridge outside the Port Authority's jurisdiction, Permut wanted to build his new crossing north of the New Jersey state line.

As envisioned by the study, the new rail line would connect with Metro-North's existing north-south routes and provide service to Grand Central Terminal and Westchester, as well as Orange County's Stewart Airport, which had the capacity to become the New York metropolitan area's fourth major airport after Kennedy, LaGuardia, and Newark.

Metro-North's study concluded that new train services would improve regional mobility, help control growth, reduce travel times to midtown Manhattan, divert cars from highways, and link affordable housing with employment centers in New York City and in the suburbs. Although the new rail line would provide commuters heading to Manhattan's East Side with faster service, passengers destined for the western part of Manhattan would still find it faster to use the existing service and transfer in New Jersey for trains to Penn Station.[8]

Permut realized that Metro-North's planning consultants could kill his dream of building a new Hudson River crossing. If they forecast low ridership for the new train service, Metro-North would not qualify for the federal funds it needed to pay for the project. The railroad needed to show that if it built a new rail line, many commuters would take a train rather than drive to work.

The consultants realized the uncertainty of their efforts.[9] They used computer models that were originally developed to estimate the number of bus riders, but commuter railroads attracted a much different demographic than buses. Suburban train riders had higher incomes and were much more likely to own a car than urban bus riders. The lead consultant remembers that their work, a combination of both art and science, was in some aspects "crystal ballish."[10]

Permut deliberately used a number of overly optimistic assumptions to boost a new rail crossing's expected ridership. According to one planner closely involved with the project, Permut "went ballistic" when he saw that the consultant's initial ridership projections were low. He kept insisting on "better numbers" and would tell his staff and the consultants to come back with higher estimates. To accommodate Permut's requests, one Metro-North planner said, "We could always come up with other factors that would raise the ridership numbers, like we could say that population would grow more than we originally thought." According to a Metro-North planner who worked on the study, the final ridership numbers Metro-North published were "close to realistic although there is a fine line between optimistic and unrealistic."[11]

Overall, Metro-North made at least six overly optimistic assumptions to increase the estimated number of commuters who would shift from cars to trains:[12]

- Trains would operate at one hundred miles per hour for most of the distance, even though it was more likely that trains would average seventy, with a maximum speed of one hundred.[13]
- Bus travel times would get 40 percent worse in the next twenty years, even though the state was expected to build an HOV lane along I-287. (In fact, bus travel times between Rockland County and Manhattan stayed about the same for that period.)[14]
- Free or low-cost vans would meet all trains in suburban stations and carry passengers to their employment destinations without any wait time.
- Employers would offer disincentives and incentives, such as limiting parking and subsidizing their employees' rail fares.
- A significant number of new trips would be induced by the new railroad; that is, people would make more trips because the railroad was built. (The consultants felt that this assumption was definitely more art than science.)[15]
- Sixty-five percent of new rail users would be diverted from cars, while only a relatively small portion of the existing transit users would shift to the new train service.

Every one of these assumptions was theoretically possible. At the end of the day, the consultants felt they had maintained their integrity and not buckled to Permut's pressure as most other consultants would have. But they realized that if they didn't bend somewhat they "could get a reputation for not giving in to a client, and that could hurt you getting more business."[16]

Permut's approach to estimating ridership was not unusual for transit agencies completing rail projects at the time. In a 1992 article titled "A Desire Named Streetcar," Don Pickrell revealed that local officials seeking federal transportation funds for new rail projects were grossly overestimating transit ridership. Nearly all the rail lines that had been built recently in the United States were carrying less than half the number of riders forecast.[17]

Just as Metro-North's ridership estimates were suspect, so too were its cost estimates. In 1994, Metro-North estimated that construction costs could be as little as $1 billion. When planning was conducted two decades later under the federal government's close scrutiny, cost estimates were about ten times higher.

Despite all of the overly optimistic assumptions, Permut's rail plan received little public support. Just as communities along the river opposed a new highway bridge, many were also against a new rail crossing.[18] In Westchester, the railroad's planners were accused of trying to create an empire, and they faced vocal opposition from residents concerned about noise, construction, and the loss of river views. In Rockland, the planners heard from residents living near the Hudson River who were still bitter about the state's decision to build the Tappan Zee Bridge through their communities.[19] Even Rockland County Executive John Grant had serious misgivings about building a new rail line between Rockland and Manhattan.[20]

Furthermore, the proposed rail crossing generated opposition from the environmental community. In 1989, George Case, president of the Federated Conservationists of Westchester County, argued that connecting a new rail line with the existing Hudson Line would be costly and an environmental disaster. Two years later, he told the MTA Board that a new group he helped form would "fight to the death" any new Hudson River crossing, rail or roadway, in the Tappan Zee area.[21]

Metro-North did not even receive much support from its own parent organization. In the early 1990s, Peter Derrick led the MTA's efforts to coordinate and compare potential transit megaprojects in the New York region. He determined that Metro-North's plan rated poorly compared to other projects in the region. At the time, New York and New Jersey transportation agencies were studying options to build a new rail line under the Hudson River between New Jersey and midtown Manhattan as part of its Access to the Region's Core project. Derrick told a Metro-North planner that it was "insane for Metro-North to be thinking" about building a new rail line across the river between Rockland and Westchester Counties when a new rail line to Manhattan would serve so many more people.[22]

In 1995, the MTA Board put Metro-North's study on hold because the project was expensive, funding was uncertain, and the proposal had not been well received. One Metro-North planner felt the only one supporting the railroad's plans was the Mid-Hudson Pattern for Progress, a business-oriented civic group based in Orange County.[23] However, its executive director, William J. D. Boyd, realized that the route across Westchester County would pose terrible environmental difficulties and go through some of the most expensive real estate in the country.[24]

When NYS DOT planners reviewed Metro-North's analysis in the mid-1990s, they found numerous flaws associated with building and operating a new rail line across Westchester County. No matter where the stations were located, a costly shuttle bus system would be needed because residential and commercial developments were not concentrated. The planners calculated that a commuter rail line across Westchester would be vastly underused: it would only need two or three train cars running every twenty minutes, whereas most commuter railroads typically run trains with more cars and greater frequency. NYS DOT's Regional Office politely reported that a rail line across Westchester would be an imprudent use of the limited funds available for transportation.[25] That meant NYS DOT's planners thought it was crazy to spend over a billion dollars for a rail line across Westchester County when a few buses could be run on an HOV lane, instead.

NYS DOT's planners were also critical of Metro-North's rail line because they were following orders. In 1989, when NYS DOT commissioner Franklin White heard that Metro-North was considering building a rail line along the Cross Westchester Expressway, he told his staff that under no circumstance should the department endorse this idea.[26] He did not want it to interfere with the state's plans to rebuild the expressway and add an HOV lane.

Despite these obstacles, Permut was able to strike a deal with NYS DOT officials. They agreed that NYS DOT would not oppose a new Hudson River rail crossing that connected with Metro-North's Hudson Line along the river's eastern shore, and, in exchange, Metro-North agreed to write in its report that ridership levels did not justify building a new rail line across Westchester County.[27] Permut's deal was helpful, but if he wanted to fulfill his vision, he would need a powerful political leader to champion the project.

Governor Pataki's Vision

In the beginning of his second year in office, Governor Pataki decided the time was right to develop and announce a vision for the future of New York's transportation system. In preparation for his May 1996 address to the New York City Building Congress, the governor's office looked at a list of megaproject ideas generated by the state's transportation agencies and authorities. The governor and his staff considered whether a project was feasible and affordable, what constituency it served, and whether that constituency helped the governor get elected. Then they considered how to prioritize the projects, package them, and give them a new "twist."[28]

In his 1996 speech, Pataki announced a multi-decade transportation program that would make "travel by transit much more appealing and convenient."[29] His two priority megaprojects would be a new train line to New York City's Kennedy Airport and a rail connection between the Long Island Rail Road and Grand Central Terminal.

Metro-North's rail crossing did not make it onto the governor's list of priorities. Likewise, there was "no excitement" in the governor's office about the Access to the Region's Core project, which involved building a new rail tunnel under the Hudson River between northern New Jersey and midtown Manhattan. A senior Pataki official who helped develop the governor's transportation vision said about the rail tunnel, "It's New Jersey, so who cares."[30] The MTA would later pursue two other megaprojects. The powerful New York State Assembly speaker, Sheldon Silver, championed the Second Avenue Subway. New York City's mayor, Michael Bloomberg, successfully obtained the funding and approvals to extend the number 7 subway line to Manhattan's far west side.

After Metro-North's rail feasibility study was published in 1995, it remained on the shelf for a few years until Howard Permut received a phone call from Albany. The Thruway Authority had a bridge it wanted to replace, and it was interested in working with the railroad.[31]

The Thruway Authority Considers a New Bridge

The Thruway Authority has long been aware of the Tappan Zee Bridge's deficiencies, going back at least to John Shafer, the Thruway Authority's executive director between 1987 and 1995. No one in New York understood the problems of the state's aging bridges better than Shafer. Before joining the Thruway Authority, he was NYS DOT's assistant commissioner and chief engineer; in that position he was responsible for the inspection and safety of thousands of highway bridges all across the state.

During his tenure, Shafer and his chief engineer at the Thruway Authority, Tony Gregory, held numerous talks about the Tappan Zee Bridge. They thought the bridge was "functionally obsolete" because it could not accommodate the growing volume of traffic and did not have shoulders along the side of the roadway to store disabled vehicles and provide access for emergency vehicles. However, they determined that the bridge could last forever as long they replaced parts when needed.[32]

Shafer and Gregory identified some major issues with the bridge. It needed stronger protection from passing ships, its deck needed to be replaced because it was "deteriorating and taking a beating," it needed to be strengthened to meet new wind-bracing requirements, and the structural steel needed to be repaired.[33]

They received unsolicited proposals to replace the Tappan Zee Bridge with a wider one. An engineering firm estimated that a new bridge would cost $1 billion. Another firm came up with a scheme for adding more lanes to the existing bridge, but the Thruway Authority engineers determined that was not feasible. To accommodate the high traffic volume, Shafer's engineers also considered building a new bridge parallel to the existing one.[34]

From Shafer's perspective, there were only two reasons to replace a bridge—either to add more capacity or to avoid spending more on rehabilitation than on replacement. Spending $1 billion for additional capacity did not make much sense to him given all the other investments the authority needed to make. The authority had 3,240 lane miles of roadway and 1,039 other bridges to maintain.[35] In 1992 it even became responsible for the money-losing 523-mile state canal system. Shafer also determined that over the long term maintaining the bridge would be less costly than building a new one.

Shafer undertook a detailed seismic investigation of the bridge.[36] Engineering consultants found that it could survive a fifty-year earthquake (one with no more than a 10 percent probability of occurrence over a period of fifty years) with no major damage. To survive a hundred-year earthquake, additional improvements would be needed.

Shafer called for a multiyear capital program to keep the bridge in good operating condition.[37] A month after he retired in 1995, a Thruway Authority committee finalized its recommendations for the bridge, developing a ten-year $187 million capital program to maintain and repair it. The committee also recommended that the authority prepare to replace the deck of the bridge between 2006 and 2010.[38] Shafer's successors had a different perspective about repairing the bridge, however.

In 1995, Shafer's long-time political advisor, Stephen Morgan, became the Thruway Authority's new executive director.[39] Although Morgan's engineers thought the authority could continue maintaining and upgrading the bridge, so it could "stand forever and forever," he believed that "the growth of the area and the reliance on vehicles and trucks was going to drive traffic numbers up so high that it didn't make sense to keep it the way it was." He knew that replacing the bridge would be expensive, but he thought "the link was too important to have it clogged." He hoped that the Thruway Authority would be able to obtain outside funding to pay for a new bridge, so it could reduce its ongoing investment in the existing structure and spend more money upgrading other parts of the Thruway system.[40]

A steady stream of financial and engineering firms came to Morgan's office with ideas for replacing and financing a new bridge. Morgan remembers that the Thruway Authority's engineers "didn't think anyone had the money or the gumption to build a new bridge." They were "good ideas, but how the hell were we going to pay for it?"[41]

Under Morgan, the Thruway Authority undertook steel repairs, deck repairs, and other improvements to the Tappan Zee Bridge. The authority also started replacing and rehabilitating the protection system for the piers located adjacent to the river's navigational channels.

After George E. Pataki was sworn in as New York's new governor, he appointed Howard Steinberg as Thruway Authority board chairman. In 1996,

Steinberg put together a plan to maintain the authority's aging roads and bridges. However, his plan was thwarted by the legislature's leaders since it relied on significant toll increases, After Governor Pataki publicly announced that he, too, was opposed to the planned toll increases, the Thruway Authority had to cut back on its planned long-term investments.[42]

Unfortunately, politicians are not rewarded for carefully maintaining their aging infrastructure; instead, voters reward elected officials who keep tolls and taxes low until the next election. That is why governors who want to improve their transportation systems set up authorities, borrow money, and seek out financial assistance from the private sector.

In 1996, the Thruway Authority hired John Platt, an Ohio transportation executive, as its new executive director. The governor's office recruited Platt because he promoted the private financing of transportation improvements, an idea that fit in well with the ideology of the new Republican governor, George Pataki.[43] Thruway Authority Chairman Steinberg called Platt a "national leader in pioneering innovative transportation finance programs."[44]

When Platt started working at the Thruway Authority, he wanted to undertake a prominent project with the private sector.[45] Public-private partnerships for financing, designing, constructing, and operating new roads and bridges were becoming an increasingly popular way for governments to build large-scale projects at the time.[46] Private companies were becoming more interested in financing and operating highways because of new cost-effective electronic toll collection systems such as E-Z Pass.

Platt would eventually focus his efforts on replacing the Tappan Zee Bridge, a span that was unable to accommodate growing traffic volume. According to Lawrence DeCosmo, the authority's chief financial officer at the time, the number-one reason the Thruway Authority wanted to replace the Tappan Zee Bridge was to add highway capacity.[47]

The seventh lane on the bridge was accommodating more vehicles, and a movable barrier system installed in 1993 helped even more. It provided drivers with four lanes eastbound in the morning and four lanes westbound in the evening. Thruway Authority officials felt additional capacity was needed in both directions, however.

Authority officials were concerned about future revenue. The Tappan Zee Bridge was an important contributor to the Thruway Authority's bottom line and helped subsidize the rest of the 641 miles of Thruway roads. Not only was the three-mile bridge providing about 16 percent of the authority's annual toll revenues in the mid-1990s, but a more congested bridge would deter drivers from using other parts of the Thruway.[48]

There were other reasons to replace the bridge besides traffic and money. Keith Giles, the Thruway Authority's chief engineer, was concerned about its seismic vulnerability and its lack of internal redundancy: critical portions of the bridge were not supported by more than one steel member. To make matters worse, the salt and deicing chemicals used on the bridge had corroded some of the steel structure.[49] Giles was also concerned that marine borers could one day eat the timber piles that supported the one-and-a-half-mile causeway portion of the bridge.

Giles felt that "heavy replacement" and ongoing maintenance could keep the bridge up, but it was becoming a big sinkhole.[50] Another Thruway Authority engineer said that the bridge had become like an old car. "You could keep on pouring more and more money into maintaining it, but it would not have collision-avoidance system, back-up cameras, or antilock brakes."[51]

Paine Webber, a New York financial firm, offered to help Platt think through the process of replacing the bridge. The Thruway Authority had retained the company to underwrite its bonds, a lucrative business in which investment bankers purchased and then sold them at a profit.

Paine Webber hosted a two-day meeting collaborative design meeting, called a charette, in May 1997 for Platt and the authority's senior staff to discuss the problems pertaining to the bridge and the key issues associated with replacing it. The Paine Webber official who organized the charette realized the Thruway Authority staff did not have sufficient experience in planning, let alone building a major infrastructure project.[52] The Thruway Authority maintained its facilities and rebuilt interchanges, but not since the Thruway was built had it taken on a project that matched the complexity and cost of replacing the Tappan Zee Bridge.

Paine Webber brought in bridge builders, bridge designers, tunnel designers, lawyers, and financial experts. The Thruway Authority received free advice; it did not pay Paine Webber for the experts' time or the materials they prepared.[53] Those came at a different type of cost: the advice was biased, since every single private-sector person in the room had an incentive to replace the bridge. The lawyers, engineers, and builders could receive lucrative contracts while Paine Webber could underwrite more bonds.

The charette's participants discussed various issues associated with replacing the bridge, including growing travel demand, federal regulations, procurement methods, and the possibility of working with other transportation agencies such as the MTA.[54] They also discussed various forms of public-private partnerships.

The charette would serve as a turning point in the history of the Tappan Zee Bridge. Although it did not end with definitive plan, it crystallized the Thruway Authority's desire to replace the bridge.

Did the Bridge Need to Be Replaced?

The need to replace the bridge in the late 1990s was not clear-cut. Platt, who died in 2004, seemed to think it made sense to build a new bridge because of capacity constraints, operational deficiencies, safety concerns, costly maintenance requirements, the need to re-deck the bridge, and the opportunity to leverage private and federal funds. The other option, rehabilitating the bridge, would have been costly as well as disruptive to the authority's customers.

Interviews with two former New York State chief engineers indicate that if NYS DOT had owned the Tappan Zee Bridge, the state probably would have chosen to repair rather than replace it. Shafer thought that the authority officials pushed for a new bridge because it wanted to add new capacity, while Robert Dennison believed that the potential to increase toll revenues was an important consideration.[55]

If NYS DOT had owned the bridge it would not have considered toll revenues, since its bridges did not have any tolls. Moreover, NYS DOT was not as keen on a wider span because that would induce more people to drive, thereby worsening traffic conditions on its already congested roads in Westchester and New York City.

Longtime Thruway Authority consultant Dan Greenbaum led the first briefing at the 1997 charette. He said, "There were all sorts of efforts to manufacture reasons why they needed a new bridge. It was not terribly convincing to me." He thought that the Thruway Authority mainly wanted to replace the bridge because "it did not meet the high level of service for which it prided itself on."[56] Although, he added, "To be fair, there were lots of reasons to replace it, including the very high cost of maintaining the bridge. However, you can do a lot of maintenance and rehabilitation for the equivalent amount of the annual debt service of a new crossing."[57]

Before the Governor's I-287 Task Force convened, Platt never talked publicly about the Thruway Authority's desire to replace the Tappan Zee Bridge. However, in private conversations, he exaggerated the need to replace the bridge and minimized the cost to replace it. He told Lou Tomson, the Thruway Authority chairman, that the bridge had to be replaced because it would not be safe to operate by 2007, or even functional by the year 2010.[58] He told other senior state officials that the cost of replacing the bridge would be about the same as the cost to repair it.

In 1997, Platt's and Permut's visions had become intertwined. Permut wanted the Thruway Authority to build a new bridge that could accommodate a rail line because Metro-North could not afford to build its own Hudson River crossing. Platt figured that a new rail line would help him generate both political support and transit funding for a new bridge. The timing was ideal for the two men since the governor, his senior officials, and the public were all focused on the I-287 congestion problem, and they seemed to support an ambitious effort to expand the rail system and alleviate bridge traffic.

6

Pataki's Task Force

Raising Expectations Sky-High (1998–2000)

Governor George Pataki's decision in 1997 to cancel plans for the HOV lane and set up a task force to study potential congestion solutions was a crucial turning point in the I-287 saga. It would add another fifteen years to a planning process that had begun nearly twenty years earlier.

Pataki surveyed a fractured political landscape filled with potential land mines. Residents and elected officials along the I-287 corridor were concerned about widening the highway, the "construction guys were teed off at him" for canceling the HOV lane, and the business community wanted to increase transportation capacity.[1] Meanwhile, many Westchester environmentalists sought a light rail line, the Tri-State Transportation Campaign leaders wanted congestion pricing, and the river communities were against any new crossing whatsoever.

Not long after canceling the HOV lane, Pataki told a *New York Times* reporter, "You can't govern without bad ideas, because no one has nothing but good ideas. But what is most common is that you don't have any ideas. I always have what I would call creative, others might call bizarre, ideas as to how we can move forward as a state."[2]

The governor wanted the task force to consider and develop his idea for the I-287 corridor. It combined the visions of Thruway Authority executive director John Platt and Metro-North vice president Howard Permut by replacing the Tappan Zee Bridge and building a new commuter rail line that would connect

Westchester, Rockland, and Orange Counties (including Stewart Airport) with Grand Central Terminal. Pataki would have considered a light rail line or new bus services if he knew commuter rail was not practical, but no one told him that it wasn't.[3]

Pataki's Strategy

Pataki often relied on advice from two of his loyal Westchester friends. Virgil Conway, the chairman of the Metropolitan Transportation Authority (MTA), was an early supporter of Pataki's gubernatorial campaign and a member of the governor's self-described "kitchen cabinet."[4] John Cahill, the state's environmental commissioner and Pataki's former law partner, was the only person the governor brought along to meet with the Tri-State Transportation Campaign leaders on the day he decided to cancel the I-287 HOV lane.

In separate interviews, all three men discussed how the I-287 Task Force needed to consider the region's long-term needs. Pataki told me, "We had a pretty good idea of what we wanted to do" before the task force even convened. He thought the HOV lane would provide only a short-term solution to a long-term problem, and he wanted his task force to prepare the lower Hudson Valley for long-term economic growth in an environmentally sustainable way. The governor said he was planning for one hundred years and he did not want to build something that would fall down or be obsolete after thirty. He clearly preferred rail to any bus option.[5] Conway, referring to the three counties north of Rockland County, said, "We need to prepare for when Orange County develops, and Sullivan and lower Ulster."[6]

Cahill said, ["We] needed to plan for tomorrow, not just for today." He explained that "there was no question that [Pataki] was leading on this. People didn't understand, but he had a vision for the region, from Orange County to New York City."[7] The governor's vision was consistent with his 1996 speech to the New York City Building Congress, which called for a multi-decade program of transit improvements to reduce highway congestion and make "travel by transit much more appealing and convenient."[8]

Pataki and his inner circle did not want to make the same mistake as their predecessors in the governor's office. Cahill said the state should have accommodated a future train service when it built the original Tappan Zee Bridge.[9] Likewise, Conway said the Tappan Zee Bridge had been built "on the cheap, with no thought of the future." He said, "It's difficult to justify spending a billion dollars [to replace the bridge] without providing added value to the region." Public transportation, Conway said, "is the only way to get people to stop driving."[10]

Pataki and Conway were plagued by decisions made during the era of Robert Moses, New York's "master builder," who between the 1920s and 1960s reshaped much of the region to accommodate the automobile. To this day, his legacy hangs over every New York leader who wants to improve the region's transportation system. Under Moses's advocacy and direction, New York took advantage of federal transportation funds and toll revenues to build an extensive network of bridges, tunnels, and highways. Since Moses did not have to worry about strict environmental regulations and extensive public participation, he was able to build roads through country estates, sensitive wetlands, thriving industrial areas, and low-income neighborhoods.

However, Moses neglected the region's existing subway and commuter railroad system, which moved millions of New Yorkers much more efficiently than automobiles. Pataki's aspirations to connect Kennedy Airport with Manhattan by train were exceedingly complicated and expensive because the highways Moses built to the airport could not easily accommodate a rail line. As MTA chairman, Conway operated the bridges and tunnels that Moses built, but none of them could be retrofitted to accommodate train service.

Conway and Pataki both had a soft spot for trains. Conway's father and uncle worked for a railroad.[11] Pataki appreciated the economic and environmental benefits of rail service and had fond memories of taking the train to major league baseball games and sleeping overnight on the benches at Grand Central Terminal when he missed the last train back to Peekskill in Westchester.[12] Pataki did not have the same feelings for buses as he did trains. He was not known for reminiscing about long bus rides or extended stays at Greyhound stations.

Pataki understood how a new east-west railroad line like the one envisioned by Permut could tie the region together. Before he was governor, Pataki often drove from his home near Metro-North's Hudson Line to downtown White Plains near the railroad's Harlem Line station. He would have preferred taking a train for that leg, but there was no east-west rail service.[13]

Pataki had a strong environmental record as governor when it came to promoting public transportation, protecting open space, and cleaning up the Hudson River. However, he was not worried about the prospect of new transportation capacity exacerbating the region's sprawl. He had faith that local communities could prevent unwanted development.[14] He thought a new rail line would promote long-term economic growth in an environmentally friendly way. Pataki's vision and his periodic conversations with Conway shaped the recommendations of the task force.

Unlike previous committees that had studied the corridor, Pataki's I-287 Task Force took its direction directly from the governor. Only his top officials served on the body, and he kept a steady eye on their progress. Pataki told the task force members that the state's transportation investment needed to be made in a way that would accommodate the region's growth.[15] Pataki publicly kept himself at a safe distance from their work, however, to avoid any potential political repercussions.

Pataki appointed his most trusted and senior advisors to the task force because he had plans for an ambitious project of statewide significance. Its members included Cahill and Conway as well as another confidant from the governor's inner circle, Charles Gargano, who had raised 14.5 million dollars for Pataki's first gubernatorial campaign and now led the state's economic development agency.[16] They were joined by Platt and Joe Boardman, commissioner of the New York State Department of Transportation (NYS DOT).

The task force set up a twelve-member advisory committee that included representatives from local governments, business groups, and environmental organizations. Many members of the advisory committee had been part of the HOV battle, including three people associated with the Tri-State Transportation Campaign (Maureen Morgan, Jim Tripp, and Jeff Zupan), the Westchester County Association's Larry Dwyer, and Ross Pepe from the construction industry. The task force members met publicly with their advisory committee between 1998 and 2000.

The task force hired a consulting team to examine the corridor's long-term needs and assess potential improvements. The consultants, who were closely supervised by Task Force chairman Virgil Conway and his senior aide Susan Kupferman, regularly presented materials to the task force and its advisory committee.

Finalizing a Plan

When Pataki canceled the HOV lane in 1997, he said that NYS DOT would address certain bottleneck issues along the six-lane Cross Westchester Expressway.[17] One of the task force's first actions was to sign off on this initiative. In June 1998, at the first meeting of the task force and its advisory committee, NYS DOT's Robert Dennison, a politically savvy engineer who had been brought in to replace Al Bauman as regional director after the HOV debacle, presented his plan.[18]

Dennison explained to the panel how NYS DOT would implement "operational and safety improvements" that involved widening ramps and reconfiguring interchanges.[19] His design incorporated suggestions that had been made by the Westchester business community and the Regional Plan Association's Zupan during the HOV debate. Referring to the head of the construction industry trade group, Dennison remembers that state officials were "trying to please Ross Pepe and others of his world."[20]

Dennison's presentation effectively ended years of debate about whether the expressway should have six lanes, eight lanes, or an HOV lane. Studies that had begun in 1978 to address safety and traffic improvements along the Cross Westchester Expressway quietly and anticlimactically came to closure.

Dennison figured out how to add capacity without engendering community opposition or a lengthy bureaucratic process. Since adding a new lane more than one mile long would have triggered federal requirements to initiate another environmental review and conduct more public hearings, Dennison had his staff design "acceleration and deceleration lanes" up to nine-tenths of a mile long instead. These lanes would in effect turn I-287 in Westchester into an eight-lane highway where traffic warranted it.[21]

NYS DOT also gave the project a great name. Widening a highway was controversial. In comparison, "operational and safety improvements" sounded modest and necessary. It was the motherhood and apple pie of highway work.

Chairman Conway cleverly avoided opposition to the highway reconstruction by co-opting the Tri-State Campaign leaders on the advisory committee. NYS DOT officials were expecting to present many more details about the project, but Conway did not want the advisory committee members and the public to focus on them.[22] Conway referred to the proposed highway work as an early action item of the task force. Instead of spending too much time discussing the proposed highway improvements, he steered the task force's discussions toward the corridor's long-term problems and asked each of the advisory committee members to identify potential solutions. The HOV opponents did not realize the magnitude of the highway reconstruction project that NYS DOT was about to undertake.

Dennison told the task force and advisory committee that the reconstruction of the expressway was expected to cost $265 million, which was much higher than the governor's previous estimate of $150 million.[23] Ultimately, those numbers would be dwarfed by the actual project cost, which would exceed $740 million.[24]

The results of the highway project would be popular, though. East-west traffic would move faster especially at the critical interchange between the Thruway and the Cross Westchester Expressway. One planning participant, real estate developer Bob Weinberg, was distressed after construction began, however. He had supported the HOV lane along the expressway median because it would have provided a dedicated right-of-way for carpoolers and buses. In the future, if ridership warranted it, the state could have built a light rail line or monorail along that right-of way. When Weinberg saw the state reconstruct the expressway's overpasses with support columns in the median, he realized the chances of the state's ever building a transit line down the middle of the expressway for buses or trains were now slim, because in order to do so it would have to reconstruct the overpasses completely all over again.

The Three-Step Planning Process

Once the task force members had taken care of this early action item, they turned their attention to the corridor's long-term needs and opportunities. Working with the MTA, NYS DOT, and Thruway Authority staff, the task force and its consultants followed a three-step process to develop a set of recommendations. First, they documented the nature of the corridor's traffic problems and projected future conditions. Second, they identified potential alternatives to solve the problems. Third, they evaluated each alternative in terms of its ability to meet the task force's goals of improving mobility, fostering growth, and minimizing environmental impacts in a cost-effective manner. This is known as a *rational planning process*, and it is the dominant paradigm for planning transportation projects.[25]

The task force's consultants undertook an extensive study of the corridor. They measured traffic conditions, reviewed available data, and conducted qualitative surveys. They focused on the peak periods of travel, which were 6:00 A.M. to 10:00 A.M. eastbound and 3:00 P.M. to 7:00 P.M. westbound. The task force, advisory committee, and the media found the consultants' work useful, because it brought updated information together about the nature, magnitude, and underlying causes of the existing conditions.[26]

The analysis showed the complexity of the congestion problem. Traffic was growing rapidly at the same time that it was spreading out over more hours and across a wider geographic area, reflecting a region-wide phenomenon of more suburb-to-suburb commuting. It also revealed a wide variety of origins and destinations. Most vehicles on I-287 were not crossing the Tappan Zee Bridge. Of those that were crossing eastbound in the morning, about 53 percent were traveling to Westchester and another 30 percent to New York City. The others were heading to Long Island, Connecticut, and destinations outside the New York metropolitan area. Reverse commutation traffic also was growing rapidly. In the 1980s, the number of morning trips from Westchester to Rockland and New Jersey grew by about 80 percent.[27]

The state and counties had made a concerted effort to promote carpooling and bus use, but it was having little effect on reducing congestion. Only 3 percent of the trips along the corridor were made by public transportation and most drivers were driving alone. The average vehicle occupancy was about 1.16, which meant that every twenty-five vehicles only carried twenty-nine people.[28]

The consulting team used the information it collected about current traffic to estimate future conditions. At the time, driving from the western portion of Rockland to the eastern end of Westchester on I-287 took thirty minutes without traffic; the same trip during the peak hour was one hour and ten minutes long. In 2020, the peak-period trip was expected to take between one-and-a-half and two hours even with NYS DOT's operational and safety improvements.[29]

The consultants provided the task force and advisory committee with information about the condition of the Tappan Zee Bridge, but they did not perform an independent cost analysis.[30] Instead, they used estimates prepared by the Thruway Authority, which indicated the cost to rehabilitate the existing span—including replacing the causeway, protecting its piers from vessel collision, retrofitting the structure for earthquake preparedness, and replacing the deck—would be $1.1 billion over thirteen years. The Thruway Authority, which was trying to avoid these expenses, claimed the cost to replace the bridge with a wider one would be only $1.2 billion.[31]

FIGURE 6.1 This photograph shows a bus traveling along a bus guideway in England. (*Source:* 2011 photograph of Cambridgeshire Guided Busway route by Bob Castle, http://commons.wikimedia.org/wiki/File:Guided_bus_opening_day2.jpg. Creative commons license.)

After identifying and documenting the nature of the problem, the next step in the planning process was identifying the various strategies that could address the corridor's long-term needs. The consultants identified a list of sixty potential options based on ideas from NYS DOT, Thruway Authority, the MTA, the advisory committee members, and employers in the corridor. These alternatives were very similar to the strategies that had been considered during the previous two decades. They included transportation demand management, congestion pricing, commuter rail, light rail, and expansion of the highway and bridge.[32]

Only one important new alternative was identified. Three politically connected Westchester businessmen bought the American rights to a guided bus technology and then lobbied local and state officials to install it along the I-287 corridor.[33] Using this technology, buses could operate along roads with and without guideways (or tracks) embedded in the roadway (see figure 6.1). The firm convinced the governor's office that guideways would allow buses to travel faster than regular buses, even though that was not true.[34]

An alternative that received a great deal of attention was congestion pricing. At the time, car drivers paid a three-dollar toll, but most peak-period traffic consisted of regular commuters who paid only one dollar if they crossed the bridge a minimum of seventeen times a month. The Thruway Authority commissioned a congestion-pricing study that showed that increasing peak period tolls would spread out traffic volumes over a longer peak period.[35]

The consulting team, under the direction of the MTA, NYS DOT, and the Thruway Authority, grouped some of the alternatives together and recommended approximately fifteen that should be eliminated from further review. These included options considered unsafe or not physically feasible, such as allowing cars to drive along the shoulder of the Cross Westchester Expressway. They also eliminated alternatives that were already under way, such as construction of a new interchange in Orange County that would encourage some drivers on the Thruway to use the Newburgh-Beacon Bridge rather than the Tappan Zee Bridge.

The consultants eliminated alternatives that would exacerbate the congestion problem such as restricting the reversible seventh lane on the Tappan Zee Bridge to buses and vanpools. In the 1980s, NYS DOT planners had accurately predicted that if the seventh lane were not initially restricted for high-occupancy vehicles, it would be politically difficult to convert it to dedicated HOV use later.

After eliminating and grouping the alternatives, the task force agreed on seven that they felt should be examined more thoroughly. As shown in table 6.1, the first three alternatives included the cost of repairing the bridge, and the last four alternatives included the cost to replace the bridge.[36]

The consultants evaluated each of the alternatives based on the goals and objectives established by the task force. Where possible, the consultants used quantifiable measures. For example, they estimated how many people would use a new rail line and how that would affect traffic and travel times. Qualitative evaluations were given to other criteria such as whether an alternative would limit sprawl or be easy to implement.

Table 6.1
Cost Estimates of I-287 Corridor Alternatives

Alternative	Capital Cost
Transportation demand management (TDM)	$1.2 billion
Congestion pricing (value pricing)	$1.2 billion
West Shore Line (Rockland to New Jersey)	$1.4 billion
Commuter rail	$4.1 billion
Light rail	$2.6 billion
Bus guideway	$1.9 billion
Widen bridge and connecting highways	$1.7 billion

SOURCE: Vollmer Associates et al., "Final Report for Long Term Needs Assessment and Alternative Analysis."

Table 6.2
Alternatives Analysis Summary

GOALS & OBJECTIVES	TDM	Value Pricing	West Shore	Commuter Rail	Light Rail	Bus Guideway	Highway/ Bridge Expansion
Improve Mobility in the I-287 Corridor							
Decrease highway travel during weekday peak periods	Slight Effect	Slight Effect	Slight Effect	Effective	Moderately Effective	Effective	No Effect
Increase public transit use	No Effect	No Effect	Slight Effect	Effective	Moderately Effective	Effective	No Effect
Accommodate growth in regional travel with no increase in congestion	Slight Effect	Slight Effect	Slight Effect	Effective	Effective	Moderately Effective	Moderately Effective
Maintain or improve safety	Slight Effect	Slight Effect	Slight Effect	Effective	Effective	Effective	Moderately Effective
Reduce through truck traffic	No Effect	No Effect	No Effect	No Effect	No Effect	No Effect	Slight Negative Effect
Minimize Environmental Impacts							
Improve air quality	Moderately Effective	Moderately Effective	Moderately Effective	Effective	Effective	Effective	Slight Effect
Limit sprawl	No Effect	No Effect	No Effect	Effective	Effective	Slight Effect	Negative Effect
Minimize impacts in Rockland and Westchester	Slight Negative Effect	Slight Negative Effect	Moderately Negative Effect	Negative Effect	Negative Effect	Negative Effect	Negative Effect
Minimize impacts on the Hudson River	Slight Negative Effect	Slight Negative Effect	Slight Negative Effect	Negative Effect	Moderately Negative Effect	Moderately Negative Effect	Moderately Negative Effect
Develop an Acceptable Corridor Wide Transportation Strategy							
Develop strategies that addresses problems throughout the corridor	Moderately Effective	Slight Negative Effect	No Effect	Moderately Effective	Moderately Effective	Moderately Effective	Effective
Develop strategies that maximize use of existing I-287 corridor facilities	Slight Effect	Slight Effect	Slight Effect	Effective	Moderately Effective	Moderately Effective	No Effect
Time Effectiveness							
Ease of implementation	Moderately Effective	Effective	Slight Effect	No Effect	No Effect	No Effect	No Effect
Develop Cost Effective Alternatives							
Capital Cost	Slight Negative Effect	Slight Negative Effect	Slight Negative Effect	Negative Effect	Negative Effect	Moderately Negative Effect	Moderately Negative Effect
Cost effectiveness	Slight Effect	Slight Effect	Slight Effect	Moderately Effective	Moderately Effective	Effective	Slight Effect
Foster Growth in Regional Employment							
Ensure that the transportation system supports regional economic growth	Slight Effect	Slight Effect	Slight Effect	Slight Effect	Slight Effect	Slight Effect	Slight Effect

Legend: Effectiveness in Meeting Goals and Objectives		
▲ = Effective		▽ = Slight Negative Effect
▲ = Moderately Effective	○ = No Effect	▼ = Moderately Negative Effect
△ = Slight Effect		▼ = Negative Effect

The consultants' analysis and their final report were heavily edited by the transportation agencies so that they conformed to the chairman's wishes. Working together, they prepared a table, similar to table 6.2, that succinctly rated each of the seven alternatives. (The original table had colored circles; this table uses black-and-white arrows instead.)[37]

Since the column for commuter rail contained the highest scores, one can understand why the task force recommended building a new commuter rail line and replacing the Tappan Zee Bridge with one that could support a rail line. However, the scores in the table and the task force's recommendations were a result of clever manipulation built on a foundation of distortion, exaggeration, and finagling.

Distortions, Exaggerations, and Finagling

The task force's recommendations were the result of many discussions between its members and the governor that were hidden from the public and the advisory committee. Pataki wanted the task force members to assess various alternatives,[38] but his preferences led the task force to distort its analysis by exaggerating the benefits of alternatives he preferred and minimizing those he did not.

For example, the planners who compared various alternatives in the 1980s and 1990s had concluded again and again that bus lanes were less costly and provided more flexibility than other transit alternatives such as light rail, commuter rail, and ferries. Building a new bus lane along I-287 might have seemed to be an obvious alternative for the task force to study, but Conway did not want to consider it because a bus lane was too similar to the HOV lane that the governor had canceled.

Even though the transportation problems and the opportunities to solve them transcended New York's borders, the task force members did not work with New Jersey to identify potential solutions. New York's senior officials seemingly had little incentive to cooperate with its neighboring state. After all, New Jersey's residents were not going to help Pataki get reelected, its companies were not adding to the state's tax base, and New Yorkers working in New Jersey paid income taxes to the State of New Jersey. It was a parochial perspective, though, because most bus and rail passengers from Rockland and Orange Counties heading to New York City were traveling along New Jersey's highways and rail lines and then crossing the Hudson River into Manhattan. Working together, the two states could have helped each other.

The task force also ignored the benefits of improving Metro-North's existing services and facilities west of the Hudson River. It did not even consider the Access to the Region's Core project, which would have provided Rockland and Orange County residents with direct train service to Manhattan via a new rail tunnel from New Jersey.

The task force considered, but minimized, the benefits of the West Shore rail line, which New Jersey Transit was considering reactivating for passenger service in New Jersey and Rockland County.[39] The task force report stated, "It is recognized that West Shore rail service would provide benefits beyond the corridor. However, those benefits were not considered relevant for purposes of this study's evaluation of alternatives that would address transportation problems within the I-287 corridor."[40]

Another reason the alternatives through New Jersey were ignored and minimized was the influence of Permut, who had begun planning a new commuter rail line ten years earlier—in large part because he wanted Metro-North to operate its own railroad and no longer rely on New Jersey Transit. A task force consultant remembers that Permut was the commuter rail expert on whom they relied. Permut, who worked in the same building as the chairman and his senior staff, participated in about a dozen meetings with the task force's consultants.[41]

Thanks to the politically connected Westchester businessmen who owned the American rights to the guided bus technology, the task force distorted its analysis of the bus guideway system, reporting that the technology offered the advantages of both rail and bus. In fact, it would have been more appropriate to say that it offered the disadvantages of both. A bus guideway system required nearly all the construction needed for both a bus lane and a rail line. Moreover, bus operators would need to install equipment on their buses to take advantage of the guideways.

Before the task force convened, NYS DOT had conducted an analysis of guided bus technology and had determined that it would be more expensive than a bus lane but no more effective.[42] The task force's report gave the guided bus alternative a favorable evaluation, however, because the governor's office had become enamored with the concept.

By far the most flagrant distortion of the planning process was related to a new commuter rail line. As the consultants were narrowing down the number of alternatives, they recommended that the I-287 commuter rail line be eliminated from further analysis, determining that it was too costly, generated relatively low ridership, and would be exceedingly difficult to construct.[43]

In 1997, when the Thruway Authority's longtime consultant, Dan Greenbaum, helped lead the authority's Tappan Zee Bridge charette, he had been skeptical about the need to replace the bridge. As a consultant to the Governor's I-287 Task Force, now he thought the commuter rail line was prohibitively expensive.[44] Ironically, the governor put together a project that Greenbaum thought was not needed with a project he thought was too expensive.

In July 1999, the task force's consultants sent a report to the transportation agencies indicating that the commuter rail alternative should be eliminated from further review because light rail could provide more frequent service, make more local stops, serve more people, and have fewer impacts than commuter rail.[45]

Given Rockland County's low-density development, the consultants found that commuter rail service was not practical. During the peak hour, fewer than seven hundred Orange and Rockland County residents were crossing the Tappan Zee Bridge and heading to Manhattan. Even if every single one of those commuters wanted to take the train, frequent service would not be practical since one Metro-North train with ten rail cars seated approximately one thousand passengers.[46]

The consultants determined that a commuter rail line would be more expensive to build and its construction would have greater environmental impacts than a light rail line because commuter rail trains could not accommodate the corridor's sharp turns or travel its steep grades as light rail trains could.

Connecting a new rail crossing with Metro-North's existing Hudson Line along the eastern shore of the Hudson River, as recommended in Permut's plan, would be problematic since it would probably require a structure separate from a new highway bridge for much of its length. Furthermore, connecting a bridge that would be more than 150 feet above the river to a rail line along the shoreline would require three miles of gradually descending tracks.[47]

After the consultants told Permut about their findings, he tried to convince them of commuter rail's advantages over light rail. He explained how a new I-287 rail line could connect with the existing commuter rail network to provide a one-seat ride between Grand Central Terminal, Stewart Airport, and Stamford, Connecticut. Commuter rail could also operate at faster speeds and support freight service.[48]

One of the consultants later said, "It doesn't take a rocket scientist to know that light rail makes more sense on the bridge given where the Hudson Line is and the [steep] grades." He had the same perspective as NYS DOT officials had in the 1980s and 1990s. He said, "Transit improvements could be done in stages; first start with bus services, then if that was popular you could run light rail."[49]

Despite these concerns about a new I-287 commuter rail line, Conway made sure that the task force report did not undermine the governor's vision. Instead of reflecting the consultant's qualms, the report indicated that "commuter rail would provide the greatest improvements in mobility corridor-wide with the flexibility to expand capacity to meet additional travel demand well into the 21st century."[50]

The task force's simple scoring scheme (from "negative effect" to "effective") enhanced commuter rail's benefits and disguised its flaws. For example, the boundaries between the different scores (for example, between "moderately effective" and "effective") were arbitrary. The task force report made the dubious claim that building a new bridge with a thirty-mile-long commuter rail line would be just as easy to implement as a new bridge with more highway lanes. Likewise, the capital cost of a $2.6 billion light rail line was given the same rating as a $4.1 billion commuter rail line. In a flagrant omission, the evaluation did not even consider operating costs, which would have put the commuter rail line at a considerable disadvantage.

The task force report made other weak claims. For example, it indicated that commuter rail would discourage sprawl. In fact, the opposite is more likely to have been true. Extending commuter rail lines and building large parking lots to serve new stations would have spurred more suburban-style residential development in Orange County and other semirural portions of the metropolitan area.

Task Force Raises False Expectations

Conway tried to craft a final set of recommendations that would please the governor as well as the diverse interests of the twelve-member advisory committee. It was not an easy job, since some advisory committee members wanted to reduce the number of vehicles on the road while others wanted to accommodate more.

The Tri-State Transportation Campaign leaders, Zupan and Tripp, advocated travel demand management measures including congestion pricing, alternative work schedules, and limits on free employee parking, but those measures faced opposition from the business and construction interests. The president of the Westchester County Chamber of Commerce, Harold Vogt, told Conway that most of the measures suggested by Zupan and Tripp would place costly burdens on businesses and make it harder for corporations to recruit employees. Vogt thought their ideas were only serving to divert attention from the need to expand highway, bridge, and rail capacity.[51] Likewise, Ross Pepe, representing the construction industry, said that congestion relief is a road solution first, while other proposals were little more than "been there, done that failed attempts."[52]

Conway resolved conflicts by including ideas from each of the advisory committee members. He wanted to "give them all a victory." He told me it was "not like they were all going to have a party, but I wanted people to feel good about themselves." He also said, "A good chairman is a mediator and consensus builder" and "I wanted them to feel part of the solution."[53]

In April 2000, the task force issued its report recommending congestion pricing and other transportation demand management measures to please people like Zupan and Tripp. The cornerstone of the task force's recommendations, however, was "the replacement of the Tappan Zee Bridge with a structure that could accommodate a new commuter rail line extending from Stewart Airport in Orange County through Rockland and Westchester counties to Port Chester near the Connecticut border."[54] Metro-North and the Thruway Authority were given joint responsibility for building the new railroad and bridge.

Overall, the task force minimized the challenges of planning, designing, building, and financing a new bridge and rail line. This raised false expectations about the project's viability and would result in the state wasting millions of dollars and more than a decade in its planning.

When the task force members announced their recommendations in April 2000, they told the public that construction of the bridge and commuter rail could begin in two or three years.[55] Eleven years later, the state was still working on the project's environmental review. Not only were the task force's recommendations overly ambitious, they may have led the Thruway Authority to defer certain bridge improvements that exacerbated the need to replace the entire structure.

The task force ignored the reason no one had built a railroad bridge over the Hudson River within seventy-five miles of Lower Manhattan. Unlike many other parts of the New York City region, the lower Hudson Valley has a series of steep hills with valleys and rivers that run north to south. The existing railroad lines were built to operate north to south so they could follow the flat terrain of the valleys. Many of the rail lines were given names that evoke that geography, such as the Pascack Valley Line, Hudson Line, and the West Shore Line. The New Haven Line runs close to the Long Island Sound and the Harlem Valley area was named after the Harlem Line.[56] Even when the railroads were the equivalent of the twenty-first-century dot-com companies, the tycoons who created them had not even dared to build an east-west railroad in Westchester.

Platt also minimized the construction impacts. He said, "I think the biggest problem we'd have during construction is curiosity, people slowing down to watch what's going on."[57] It wasn't until twelve years later that the public would be told the details about how replacing the bridge would require taking private property as well as years of impacts related to traffic, noise, air quality, and river obstructions.

The factor most responsible for generating unreasonably high expectations was Conway's decision to completely ignore how the state could obtain billions of dollars to replace the bridge and build a new rail line. NYS DOT's HOV project was estimated to cost about $365 million, but it had a $200 million federal earmark. The task force proposed a $4.1 billion bridge and rail project without identifying any funding source, and no suggestions about funds were forthcoming from the MTA and the Thruway Authority. When asked about funding sources, their representatives simply pointed to each other.[58]

In 2000, Virgil Conway said the money to finance the new bridge and commuter rail line was available.[59] In 2011, he told me the state could simply issue more bonds to pay for it.[60] His answer reflected the way he managed the MTA. The same week that the Governor's I-287 Task Force issued its final report, the MTA was preparing to issue the largest sale of municipal bonds in U.S. history.[61] Conway's MTA debt restructuring program in 2000 led to short-term savings, but its existing debt, which would have been retired by 2013, was extended until 2032.[62] Just as the Thruway Authority used its bond underwriter's advice in 1997 when it considered whether to replace its bridge, the MTA used its own underwriter to develop its financing plan.[63] Thanks to Conway, the MTA's debt doubled from $12 billion in 1999 to $24 billion in 2008.[64]

The competition for obtaining federal funds for new transit lines was fierce, and the I-287 project was not well positioned to receive the funds. When an aide to U.S. Senator Daniel Patrick Moynihan heard about the task force recommendations he said, "There's not an endless supply of money and four billion dollars is a big number. Right now, my gut tells me that there are too many mega-projects on the table at the same time . . . all I can say is that there's about one or two dollars for every twenty dollars of priorities."[65] Likewise, the Tri-State Transportation Campaign calculated that the new rail line would not be considered cost-effective under the evaluation criteria the federal government used to make its funding decisions for transit projects.[66]

In my interview with Pataki, he noted that "there was enormous private sector interest" in the megaproject.[67] In fact, many investment funds were interested in financing and building the bridge because it was seen as a safe investment. Tolls on the Tappan Zee Bridge generated well over $100 million a year, and selling bonds was a lucrative business. However, no private sector firm was interested in building a rail line unless it received ongoing subsidies because operating costs for train services far exceed passenger revenue.

Although some public transportation advocates were skeptical about the task force's recommendations, most public reaction was favorable, since the task force appeared to carefully balance economic growth with environmental sensitivity. The recommendations faced minimal opposition because it ignored the need to raise tolls, increase taxes, or reprioritize the region's other megaprojects such as the Second Avenue Subway.

As the task force was wrapping up its work, Maureen Morgan organized a two-day conference, hosted by the Federated Conservationists of Westchester County, that helped solidify support for replacing the bridge. Called Meeting of the Minds, it brought together elected officials, planners, and activists—parties that previously had been skeptical about the need for a new bridge.[68]

Keith Giles, the Thruway Authority's chief engineer, explained to the conference participants how the bridge was deteriorating and that repair costs could be as high as $1.2 billion. Before Giles first walked into the room, someone had written "No New Tappan Zee Bridge" on a board. After Giles's presentation, the participants separated into groups. Upon reconvening, someone deleted the word "No." It now read "New Tappan Zee Bridge."[69] Giles had convinced most of the attendees that it made sense to consider replacing the aging bridge with a new one that included a transit right-of-way, bicycle lane, pedestrian walkway, and breakdown lanes. The *Journal News* coverage of the conference helped generate even more support for replacing the bridge.[70]

The advisory committee members representing the business community were pleased the state planned on adding new transportation capacity. Ross Pepe said that once you get away from the river communities where there is "a strong resentment and opposition for the fact that the bridge is even located in their community, people begin to use more *rational* judgment and they are a little bit more informed because they know that the transportation system is important."[71]

Some local officials, whom Pepe apparently considered irrational, requested that an independent study determine whether replacing the bridge was essential and cost-effective. Such a review was never undertaken. One advisory committee member, Nyack mayor Terry Hekker, thought the task force always had intended to recommend a new bridge and never really wanted to hear anyone else's ideas. She said the advisory committee was "like a drunk hanging on a lamppost. It was more for support than illumination."[72]

Some advocates and public officials were concerned the state would build a new bridge but would not have enough money for the transit component. The Regional Plan Association's Zupan said since the MTA was saddled by major capital projects, the project might be remembered as more of a highway-widening project than a commuter rail one.[73] The Tri-State Transportation Campaign's Jon Orcutt said, "To us, it's a highway and bridge project and they're dangling transit as the bait, but the transit is not real."[74] Likewise, Rockland County Executive Scott Vanderhoef wanted a guarantee that a new bridge would include transit.[75]

Those concerned about the transit component would have been even more worried if they knew Thruway Authority Chairman Lou Tomson was skeptical that a commuter rail line would ever be built. He did not think the number of riders warranted a multibillion-dollar investment.[76]

Further Delays

So why did Governor Pataki and the task force decide to enthusiastically pursue an overly ambitious project while distorting and ignoring key factors that would impact its outcome? The dynamic was self-interest fueled by a combination of optimism, ideology, and ignorance.

Pataki's innate optimistic nature was strengthened by his fortuitous election victories. Despite long odds in several of his races, he had never lost an election. In 1992, he beat a seven-term incumbent state senator by fifty-eight votes in the Republican primary. It was an impressive achievement: no incumbent Republican state senator had lost a primary race in the previous fifty years.[77] In his 1994 gubernatorial campaign, Pataki was a long shot, a relatively unknown candidate in both the Republican primary and general election campaigns, who ultimately defeated the three-term incumbent governor and political icon, Mario Cuomo. One of Pataki's top appointees said that his boss exhibited a willingness to go against the conventional political wisdom and that he was a lucky guy.[78]

Conway and Pataki were both confident about their transportation acumen because they had previously contradicted expert advice. Conway's proudest accomplishment at the MTA was introducing free transfers between the city's bus and subway system on July 4, 1997. Conway called the two-fare zones in New York City the "worst tax on the working man." Instead of listening to the so-called experts who told him it would be too expensive, he said, "I took a leap of faith. My instinct said it wouldn't cost so much."[79] Likewise, Pataki stated, "Everyone said it would cost a fortune" and "All the experts said you can't do it." He claimed, "The MTA had a massive surge in revenue. More ridership and it cost virtually nothing. It was an enormous success; a huge accomplishment."[80] Although the MTA took a financial hit after introducing free transfers and multi-ride discounts, Conway and Pataki's decision did result in dramatic increases in bus and subway ridership.[81]

Pataki did not seek out information about the commuter rail alternative that contradicted his vision. As governor, he tended to stay away from the details of programs. When I interviewed Pataki in December 2011, it was clear that he was not familiar with the enormous challenges of building a new east-west commuter rail line. The combination of Pataki's optimistic nature and his ignorance about the project's details fed into his belief that a public-private partnership could finance the multibillion-dollar project. One advisory committee member told me he thought Pataki acquired ideas but never became disillusioned with them even when facts suggested he should be.

Self-interest also came into play: the governor's decisions benefited him politically. Canceling the HOV lane was a popular choice, and setting up the task force bought him some time. Lou Tomson, who was Pataki's close advisor as well as the Thruway Authority chairman, said, "The state sets up a task force when it wants to postpone a decision. A task force recommendation delays things again, but it seems like you're taking action."[82] Pataki's strategy of distancing himself from the task force helped him politically because it buffered him from the project's skeptics.

Bureaucrats try to please their superiors when it serves their interests, and Conway perceived his role as working to fulfill the governor's vision.[83] After Pataki appointed him as MTA chairman, Conway said, "I don't see any need to separate myself from the governor. . . . I am a great believer in what he stands for. I plan to be shoulder-to-shoulder with him in carrying out his mandates." Conway said to me about the task force, "I wanted to make the governor look good."[84]

Pataki had the utmost faith in Conway and the other people reporting to him.[85] His closest aides were intensely loyal. But one advisory committee member saw this as a drawback. He thought Pataki had surrounded himself with people who were prepared to be sycophants, stating, "There was no one around him to say that you are wrong, like the emperor who had no clothes."

Trying to please the boss was repeated up and down the organization chart. The chairman wanted to please the governor. The chairman's senior aide at the MTA, Susan Kupferman, wanted to please the chairman. She knew her job depended on conforming the consultants' analysis with the chairman's desire for a new bridge and commuter rail line. In turn, Kupferman's staff and the task force's consultants wanted to please her.[86] Kupferman did not have to look too far to understand the consequences of disappointing the governor's office. When the MTA was instructed to sell valuable real estate to help a very important Pataki supporter (the owner of the *New York Post*), one of Kupferman's colleagues was fired for criticizing the decision in the presence of a senior Pataki aide.[87]

Self-confidence, optimism, and ignorance. The desire to please. It all led the task force to create false expectations that would cause many more planning delays in the months and years to come.

7

The Thruway Authority versus Metro-North (2000–2006)

When Governor George Pataki's task force in April 2000 recommended replacing the Tappan Zee Bridge and building a new commuter rail line, its members hoped that the Thruway Authority and the Metropolitan Transportation Authority's Metro-North Railroad would begin construction within two or three years.[1] That turned out to be wishful thinking.

After twenty years of studies, the state would have to restart its planning process from scratch. In order to be eligible for federal funds, the two state transportation agencies would need to follow federal planning and environmental regulations that included analyzing every practical alternative and then preparing an environmental impact statement (EIS). Metro-North and the Thruway Authority faced an even greater challenge, since finalizing their plans would require a high degree of teamwork and cooperation in the face of difficult personalities and clashing institutional interests.

The Dysfunctional Team

When the Metropolitan Transportation Authority (MTA) formed Metro-North in 1983 from the remnants of private railroads, Howard Permut was one of its very first employees. His professional planning perspective and temperament helped him overcome the attitudes of the longtime employees of the private railroads who were resistant to change. To improve services and attract new riders, he developed plans to increase the fleet size, modify the fare structure, revise train schedules, expand parking lots, and build new railroad sidings for trains to pass each other.

Permut's entrepreneurial spirit helped Metro-North increase its ridership more than 70 percent from 1984 to 2013. He identified and then grew distinct travel market segments. For example, to help New York City residents gain access to jobs along the I-287 corridor, he started new reverse peak services and worked with Westchester County officials so that buses would take Metro-North's passengers to suburban office parks. He increased ridership by adding parking, and when the parking lots could not accommodate any more cars, he started new bus and ferry services to bring people to the train stations.

In my interviews with numerous transportation officials, Permut was referred to as a "genius," "extremely smart," "very smart," "knowledgeable," "oftentimes the smartest man in the room," and a "man with a mathematical mind." He was known as a "tough negotiator," a "wheeler/dealer," and "someone who could get things done." One Thruway Authority official noticed he "has an obvious love for the railroad." An MTA official said Permut "was visionary, but he stepped on toes. Howard is very astute and single-minded, but also obnoxious and annoying."[2]

Numerous people said Permut would browbeat others in order to get them behind his agenda. Referring to Permut's personality, a senior NYS DOT official said, "You're always aware he's in the room" and "those who work for him ignore his direction at their peril." Metro-North planners and engineers who worked for Permut said he effectively "gets his way with people." He is diplomatic to the people on the outside but bullies his own staff.

The pressure Permut put on his staff and consultants in the early 1990s to generate higher ridership numbers for a new rail line across the Hudson was not an isolated incident. One planner said, "He knows how to get consultants to give him the answers he wants." Another of his former employees said that "he wouldn't tell me to lie, but he would say how we can we get to a certain number." In other words, "Come back to me with a number I like, but don't tell me how you got there." One planner said, "He will say or do anything to get everybody out of his way," while another said, "He thinks he can lie and they won't notice."

Someone noted, "He knows how to twist numbers to make them more appropriate." A transit advocate called him "sly," one of his top aides called him "sneaky," while another called him a "conniver." A former MTA official said he would "twist any fact and suck up to anybody to get to be president and build up his empire. But he got things done, that's why he moved up."

Permut's arrogance turned off many people. One official said, "He really knows his stuff and if you say something off base, he'll flick you away with his finger." A former MTA official called him "tenacious and a pain in the ass who doesn't know when to compromise." Another said he did not back down even when it was obvious that a connection from a new bridge to the Hudson Line was "too prohibitive given the grades off the bridge."

Permut is a micromanager who puts a lot of pressure on himself. Two people who worked for him for many years said that when you gave him a document to review, it would often come back with more marked-up comments than typed-up text. He would have his staff produce draft after draft until he was satisfied with the final results. One staffer explained, "He thinks everybody is stupider than him."

The Thruway Authority and Metro-North each assigned one person to manage the I-287/Tappan Zee Bridge corridor study and environmental review. Permut hired Janet Mainiero, while the Thruway Authority's executive director, John Platt, brought in Peter Melewski.

Mainiero remembers that Permut gave her two marching orders: "Thou shall not spend too much money on planning," and "Thou shall not have two crossings."[3] Permut was cautious about the planning expenses because if the state did not build a new commuter rail line, he did not want Metro-North to spend millions of dollars helping the Thruway Authority plan its new bridge. Keeping the railroad and the highway on the same Hudson River crossing was important to Permut because Metro-North could not afford to build a new one on its own.

Platt set up a new department and gave Melewski four staff members to help manage consultants, review documents, attend meetings, and coordinate issues with Metro-North. In contrast, Permut initially did not provide Mainiero with any staff whatsoever. Permut also put her in the unenviable position of having to make unrealistic demands without the ability to make any decisions. While Platt gave Melewski room to be innovative and run with the project, Permut kept Mainiero on a short leash and often tugged it tight.

Not surprisingly, the relationship between Mainiero and Melewski got off to a bad start and got worse over time. The two project managers argued about the work the consultants had to perform, how many hours the consultants would need, and how much money it would cost. When consulting teams submitted proposals to the two agencies, the Thruway Authority asked them to reduce their proposed budgets by a few percentage points while Permut told Mainiero to slash them in half.

Platt and Permut had very different expectations for their project managers. Platt wanted Melewski to get the study off the ground quickly while Permut expected Mainiero to attend meetings, take diligent notes, and report back to him. Mainiero was seen by the Thruway Authority as an obstructionist because she was not allowed to make any decisions. Melewski remembers, "We would have two-hour meetings with Mainiero and the consultants. Then we would have to have a separate meeting with Permut."[4]

Excerpts from e-mails Melewski sent to Mainiero over the course of several weeks in the fall of 2001 foreshadowed future trouble. Melewski wrote:

> I think we can work very well together, but I have to tell you that a continued obsession with control (and not on the project) will seriously impact the project and our working relationship.[5]

> I'd appreciate it if we could get into a routine where MTA/Metro-North provide their comments at the same time as everyone else. Instead of waiting for everyone else to sign off, and then totally rewriting the document. We lose precious time and energy on re-reviews.[6]

> I'm still trying to figure out your agency's position, other than saying no to every new idea or concept.[7]

> It is impossible, impractical, and gross micromanagement for you to be involved in all meetings at all levels. There are too many things going on for one person to be at all of them . . . without seriously disrupting the schedule.[8]

The relationship between the two agencies did not substantially improve with Melewski's successors. In interviews with me, MTA and Thruway Authority officials described their counterparts with such terms as "bully," "misogynist," "incompetent," "control freak," and "liar." On conference calls, Metro-North and Thruway Authority staff would get into verbal fights, curse, and hang up on each other. Often the arguments were not even about the substance of the project but rather about which consultant would perform specific work, and which staff on the consultants' teams should be used.[9]

An MTA planner who went to just one meeting with the two agencies was surprised to find "a poisonous environment." He told me, "These people really don't like each other," and said that the attendees were "nasty, negative, and accusatory with everyone challenging each other."

When Melewski first picked up the I-287 Task Force report, he realized that its authors had "cooked the books" for commuter rail. It had never occurred to him that a task force and its consultants would distort facts in order to arrive at a preconceived conclusion. Melewski wanted to conduct a transparent planning process, but he felt Metro-North only wanted to give the appearance of being up front.[10]

Transparency was important to Melewski and Platt because they wanted to make sure the Thruway Authority did not get bogged down for ten years trying to complete an environmental impact statement. They had learned from federal officials that extensive public participation would help generate community support and funding, as well as reduce the likelihood that a lawsuit would stop the project.[11] Melewski's attempts to be expeditious and transparent would ultimately shorten his tenure as the Thruway Authority's project manager.

When the two agencies started working together, they agreed that all reports and press releases relating to the study would be approved in advance by the

Thruway Authority, Metro-North, and the MTA's headquarters. Melewski could get quick sign-off at the Thruway Authority since it only had about three thousand employees and replacing the Tappan Zee Bridge was one of its top priorities. The MTA, however, had about sixty thousand employees and it was much more concerned about issues relating to its own existing facilities.

Melewski was sometimes a little too eager to keep the planning process moving along. When he felt that the MTA and Metro-North were reviewing documents too slowly, he sometimes said, "The hell with it, I'll just issue the press release."[12] That led to confrontations between the highway and transit agencies.

One of those confrontations started with a line on a map. Before the first round of public meetings, the consultants prepared a poster board with a map of the existing rail lines, similar to the one shown in figure 7.1. The map included the West Shore Line, a north-south rail freight line that had once been used for passenger services. At the time, New Jersey Transit was interested in reactivating passenger railroad service on that line between Rockland and northern New Jersey.

Permut saw the West Shore Line as a threat to Metro-North. If New Jersey used it for new passenger services, Metro-North would be less likely to obtain funding for its proposed east-west rail line. When Permut insisted that the West Shore Line not be shown to the public, Melewski responded, "You can't pretend it doesn't exist." Melewski felt that Metro-North was more concerned about New Jersey Transit encroaching on its territory than in actually building something.[13] The conflict about a single line on a map resulted in a meeting at the governor's office at which Melewski was chastised for producing a poster board that had not been approved by all the parties. The West Shore Line ultimately was removed from the map and the Thruway Authority found another project manager.

The Thruway Authority's senior executives typically backed down from fights with Metro-North. Permut was more persuasive, and the Thruway Authority's leadership felt they did not have the muscle to compete with an MTA agency. One senior Thruway official characterized the MTA as "huge and powerful." Another Thruway official said the MTA had greater access to the governor and spoke for the entire metropolitan area. According to the Thruway Authority's deputy executive director, Leonard DePrima, saying no to the MTA would have "just ended up creating aggravation and animosity."[14]

One of the Thruway Authority's project managers, Mark Herbst, remembers that "Howard Permut was driving the bus. He is sharp and shrewd and got his way. I respect that he fought for his interests."[15] A consultant said the Thruway Authority did not have any experience with preparing environmental impact statements, making it possible for Permut to outmaneuver them. "Permut would talk about travel demand modeling and other things that the Thruway couldn't understand. He always got his way."

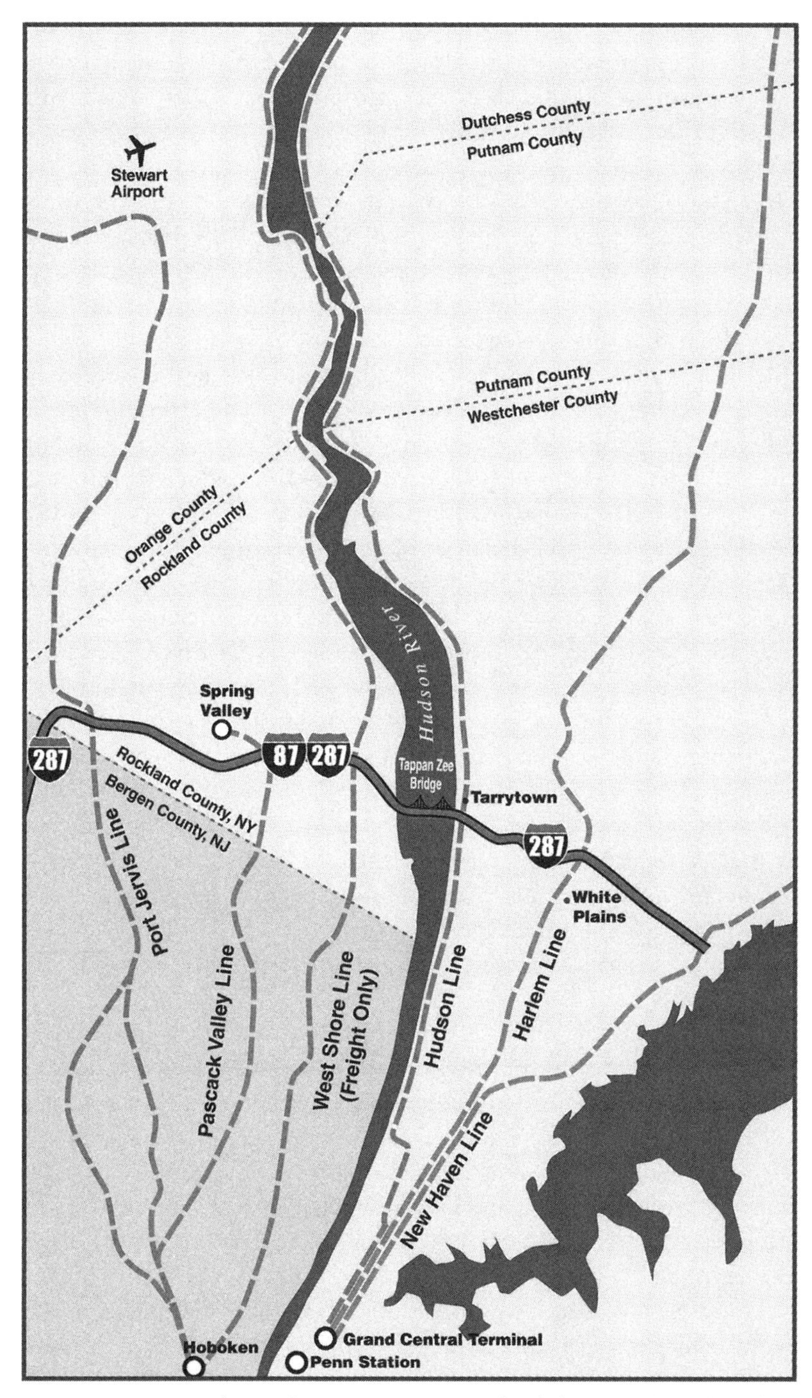

FIGURE 7.1 Metro-North provides passenger service on five different north-south rail lines in New York and New Jersey. The West Shore Line is only used by freight trains.

Although former MTA chairman Virgil Conway retired from the MTA Board in 2001, Metro-North officials still reached out to him from time to time when they had a problem with the Thruway Authority.[16] Conway was highly respected by the governor and the governor's staff, and the Thruway Authority knew that.

When trying to explain the reason for the conflict and delays, Conway told me that "highway people are highway people, and transit people are transit people; they don't mix."[17] Nevertheless, the problem was much bigger than personalities and their backgrounds. The two agencies and the people working for them had very different goals and timetables.

Melewski realized early on that the study "was a huge deal to the Thruway Authority and not as big a deal to Metro-North." A Metro-North planner told him, "If the study takes seven or eight years, so be it."[18] The Thruway Authority wanted to replace the bridge expeditiously so it could avoid spending hundreds of millions of dollars on repairs. Metro-North, however, did not provide any service along the east-west corridor and did not feel the same time pressure.

When there were backups and delays on the bridge, the Thruway Authority would hear complaints from travelers, the media, and elected officials. In contrast, Metro-North benefited from slower travel times for drivers. Heavy traffic on the Thruway made the new rail line more attractive and increased Metro-North's chances of obtaining political support and federal transit funds.

Even though the MTA bureaucracy was moving slowly, the new rail line over the Hudson River was still Permut's "Holy Grail."[19] From the Thruway Authority's perspective, however, the Holy Grail was cursed. Initially, Platt had found new train service appealing because he thought it would help him generate more support and funding to replace the bridge, but as the planning process got under way, the rail line just dragged down his efforts. The Thruway Authority's Chris Waite referred to the rail component as the Thruway Authority's "ball and chain."[20]

The railroad was literally invading the Thruway Authority's space. Building train tracks along the Thruway would take away valuable property, making it harder to maintain the roadway and service broken-down vehicles. Mainiero referred to the Tappan Zee Bridge as the Thruway Authority's "jewel in their crown." She said, "We're telling them the bridge should be this big and this robust, and we want stations and parking along their right of way. I explained to people at Metro-North that it would be like if they asked us to put a highway above our tracks."[21]

The project team, made up of agency officials and consultants, was unwieldy, beset by conflicting agendas and numerous moving parts. Between 2000 and 2006, four different Thruway Authority officials managed the project.[22] Two federal agencies, four different consulting teams, and two outspoken county executives also had their hands in the planning process. Michael Fleischer replaced John Platt in 2003 as the Thruway Authority executive director. Meanwhile, Permut's expertise, passion, and consistent presence made him a dominant player.

One of the first problems the two agencies encountered was hiring consultants. They hired four different consulting firms to manage the study, conduct public outreach, evaluate alternatives, prepare the EIS, and provide engineering support. They also had two different firms providing them with legal advice. The Thruway Authority thought having multiple firms would expedite the process.[23] Negotiations with them proved to be extensive and time consuming, however, in part because there was a lack of clarity as to where the duties of one consultant's work stopped and another began.[24] The problem only got worse after they signed contracts and began working. In retrospect, one MTA official thought that having four consulting firms instead of just one was "unworkable" and wasted three to four years.

The environmental consultants were caught between the two feuding agencies. The head of NYS DOT's Hudson Valley office, Robert Dennison, noticed that the agencies had very different perspectives on the environmental process and the role of the consultants. The Thruway Authority was not sensitive to the environmental regulations, but it listened to its consultants. On the other hand, Permut understood the importance of the environmental regulations but treated the consultants like an extension of his staff.[25] One consultant remembered, "It was a difficult project with a lot to do, only made worse because the agencies couldn't work together." The consultants tended to listen to the Thruway Authority, since it held their contracts. That infuriated Permut and supported his belief that the consultants were not objective.

Permut's marching order to keep planning expenses to a minimum added to the tension. Since Metro-North was supposed to pay half the consultant costs, Mainiero's intense scrutiny strained her relationship with both the Thruway Authority and the firms they had hired. Metro-North even refused to reimburse the Thruway Authority for hundreds of thousands of dollars of consultants' work that Metro-North claimed had not been authorized by both agencies.

Metro-North's files on the I-287/Tappan Zee Study are filled with highly critical accounts of the consultants' work. Permut thought they demonstrated favoritism toward the Thruway Authority and trivialized issues important to Metro-North.[26] He complained that their work was poorly performed and rife with significant errors. Meanwhile, the Thruway Authority thought the consultants performed "in a competent, professional, and timely manner throughout the study."[27]

According to two different sources, Permut made the relationship even worse after he pressured subconsultants to provide analysis favorable to Metro-North's interest. He told them something to the effect of "We noticed that you have a lot of work with Metro-North and we hope you can continue that." Just like

Metro-North's consultants in the early 1990s, the project team's consultants in the early 2000s felt squeezed between trying to maintain their professional integrity and pleasing Permut.

More Players Complicate Matters

Even though the I-287/Tappan Zee project was managed by New York State agencies, the federal government was legally responsible for the environmental process. Since the initiative included a highway and transit component, both the Federal Highway Administration and the Federal Transit Administration were involved. The two agencies' representatives had different perspectives, backgrounds, and lifestyles. The Federal Transit Administration, like Metro-North, was staffed with planners who worked in Manhattan. The Federal Highway Administration, like the Thruway Authority, hired highway engineers who worked in suburban Albany.

Before the two agencies even started working together, the Federal Highway Administration's district engineer, Doug Conlan, realized that combining two projects would be a mistake. He thought a bus lane on the bridge made more sense than a rail line and that working with the Federal Transit Administration would complicate the process. Also, since federal highway funds were being used to rebuild the Cross Westchester Expressway, he did not want the new highway work to be ripped up for a new rail line.[28]

Although both federal agencies had numerous regulations that needed to be followed, they treated the project differently. Since the Federal Transit Administration awarded funds to new rail lines based on their cost effectiveness, federal transit officials needed to methodically review the state's assumptions and analysis. Federal highway officials, on the other hand, strongly supported I-287 corridor improvements because they addressed problems on a critical portion of the nation's interstate highway system.

The federal agencies' most important role was to ensure that the state followed the relevant federal laws and regulations. They had conflicting interpretations and heated discussions about the best approach to conducting an environmental review, however.[29] From Metro-North's and the Thruway Authority's perspectives, trying to align the two federal agencies' procedures delayed the study by six months at the beginning of the process.[30]

As the study moved along, familiar faces tried to influence the process and outcome. In the mid-1990s, Janine Bauer from the Tri-State Transportation Campaign convinced local Westchester officials that the HOV lane was detrimental to their interests. In 2001, she told Westchester County Executive Andrew Spano how he could become a more important player in the planning process.

Bauer was concerned the state would have enough money for a wider bridge, but not enough for the transit component. She encouraged Spano to get more involved by telling him that the Thruway Authority's proposal to build a new bridge would have significant consequences for Westchester County. She described how it could increase congestion, hinder the ability of Westchester County residents to get to work, and exacerbate noise and air pollution problems for residents living along Westchester's highways.[31]

The two county executives were both voting members on the New York Metropolitan Transportation Council, the region's metropolitan planning organization, which needed to approve the use of all federal transportation funds. Since the council's rules required unanimous support from its affected members, Bauer suggested to Spano that he negotiate a set of conditions before he approved the use of federal funds to conduct the study. She pointed out that New York City mayor Rudolph Giuliani had used his veto threat regarding an issue over the JFK AirTrain project, and that Rockland County Executive Scott Vanderhoef had used the same strategy to pressure the state to participate in a study of the West Shore rail line.[32]

After Spano raised a number of issues to the transportation agencies, the Metropolitan Transportation Council's vote on the study was delayed from May 2001 until January 2002.[33] To accommodate Spano, the state agreed to set up an "Inter-Metropolitan Planning Organization" that would include representatives from four Hudson Valley counties—Rockland, Westchester, Orange, and Putnam. Both Metro-North and the Thruway Authority agreed to regularly update this new group and allow it to sign off on the alternatives to be studied.

Bringing in the county executives created much greater sensitivity to local and neighborhood needs. The political environment was such that the mayor of Nyack, for example, could not get the governor on the phone to talk about her ideas and concerns, but she could get the ear of the Rockland County executive.

Evaluating 156 Different Alternatives

The federal environmental review process required the state's agencies to follow a methodical process. In December 2002, the two agencies took the first step. They issued a "Notice of Intent" that formally announced their intention to prepare an alternatives analysis and an EIS for the I-287 corridor.[34] This notice kicked off a "scoping process" that provided the public and government agencies with an opportunity to identify potential alternatives and to comment on the process that would be used to evaluate alternatives and identify potential environment impacts.

The project team needed to conduct an analysis of all reasonable alternatives. It was very similar to the process the Governor's I-287 Task Force had followed previously. The team would start with a long list and then winnow the possibilities down to one or more alternatives, whose environmental impacts would be documented in an EIS. After issuing a Notice of Intent, it typically takes about three and a half years for the sponsor of a federally funded transportation project to prepare a draft EIS document, solicit public comments, finalize the review, and complete the process.[35] The I-287 megaproject was certainly not a typical project.

As part of the scoping process, the agencies held three public meetings attended by a total of more than 250 people in January 2003.[36] Metro-North's Mainiero heard three important things at these forums: people wanted something done about congestion, they wanted more mass transit, and they wanted to be involved."[37] All three were problematic, though. There was not much that could be done to solve the congestion problem, mass transit was expensive, and involving people slowed down the process.

The meetings revealed numerous conflicting interests. For example, some people wanted a wider highway to foster new development, while others were more concerned about sprawl. Several public speakers wanted the state agencies to select the best project for the region, whether that meant a new rail tunnel between New Jersey and midtown Manhattan or a new bridge at the Tappan Zee. Other speakers, like Ross Pepe who represented Hudson Valley contractors, were not interested in promoting projects that would be built in other parts of the New York region.

Mainiero felt that the public seemed to expect that the project would solve every transportation and environmental issue in the region. For example, the Empire State Passengers Association and others thought the new bridge could solve a problem that had long made shipping goods into New York City by train both time consuming and expensive. Nearly all container ships coming into New York harbor actually arrive into New Jersey ports.[38] If a shipper wants to put a container on a train to New York City, the train has to travel more than 140 miles north to cross the Hudson River and then travel 140 miles back south to the city. Rail advocates wanted the new Tappan Zee Bridge to accommodate freight trains so that more goods destined for New York City would travel by rail and fewer trucks would clog its highways.

Newspapers raised the public's expectations about the feasibility of various options. The *New York Times* wrote glowingly about a Westchester businessman named Alexander Saunders who promoted a seven-mile highway-and-rail tunnel between Westchester and Rockland to replace the three-mile bridge.[39] The tunnel was seen as a way to improve air quality, reduce noise, and improve the quality of life for riverfront residents and visitors. The article helped Saunders build support and credibility from community groups and elected officials. The *Journal News* gave Saunders's plan extensive coverage and referred to him as a visionary.[40]

The highway tunnel idea gained momentum and was given greater credence because the agencies were not allowed to rule out any idea prematurely. The regulatory process did not permit project sponsors to prejudge the conclusion of an environmental analysis.[41] Even though the project team quickly realized that a highway tunnel would be more expensive, more environmentally destructive, and would eliminate connections to the north-south highways near the river, they needed to complete a thorough evaluation. In the meantime, the idea received more and more favorable media attention.

I asked *Journal News* transportation reporter Greg Clary why his publication gave Saunders so much coverage. Clary said that Saunders went to every meeting and "he worked me and the *Times*." Clary thought that none of the state agencies knew anything about tunnels, and he felt it was an important idea that needed to be part of the conversation. When I asked Clary whether he might have raised false expectations about a tunnel, he responded, "You have to fill the paper; [so] do you just blow him off? He was a voice in the wilderness and you want to give people information about the options."[42]

The state itself raised expectations by avoiding discussions about how to pay for the project. In 1995, the Thruway Authority's Leonard DePrima had worried about the cost of maintaining an HOV lane along I-287. In 2002, when the community suggested ideas and solutions that would cost billions of dollars, he did not understand who would pay for them. When I interviewed him he asked, "Do we charge Thruway Authority users in Buffalo to help pay for the bridge?" He said, "No one was equipped to pay for buses, let alone rail. There were lots of good ideas, but no money."[43]

The project team gave deliberately evasive answers about paying for the project. In October 2002, Metro-North president Peter Cannito said, "We do need to become a little more creative in the sources of funding." He added, "I think there is an effort to do that."[44] In 2003, the Thruway's project manager, Chris Waite, said he was not too worried about obtaining funds. He claimed, "When you have the right project, there will be money."[45]

Maureen Morgan, the former president of the Westchester Federated Conservationists of Westchester County, was an outspoken advocate for connecting all five Metro-North rail lines.[46] She did not appreciate the extraordinary challenges of building a new east-west line and then enticing Hudson Valley residents to use public transportation for their trips to and from suburban destinations.

Rockland County Executive Scott Vanderhoef did his part to raise expectations by offering the most ambitious plan of all. He wanted to combine everyone's ideas, suggesting that the state rehabilitate the bridge and build a new highway tunnel that could carry both light rail and commuter rail trains.[47]

In March 2003, Mark Kulewicz of the Automobile Club of New York suggested that the agencies consider a bus rapid transit system.[48] By that summer, the idea had been given much more prominence, including a front-page story in the *Journal News*.[49] Bus rapid transit typically includes measures to speed up bus services and increase their appeal, such as providing buses with their own lanes, having passengers pay for tickets before they board, prioritizing traffic signals, and building attractive station areas.

In the 1980s and 1990s, various transportation planning terms and strategies were trendy, such as transportation demand management, transportation system management, HOV lanes, and light rail. Now it was bus rapid transit's turn. The federal government was promoting the concept, with numerous cities such as Boston, Charlotte, Denver, Honolulu, Houston, Los Angeles, Miami, and Orlando implementing new systems. Advocates for bus rapid transit cited its advantages over rail, including lower costs, easier implementation, and the ability to travel on city streets and directly to office parks.

Buses faced an uphill battle, however. In 2003, the owner of the private bus company that operated Westchester County's system was asked by a *Journal News* reporter why he did not speak out in favor of bus rapid transit. He laughed and said, "Have you ever tried talking in a wind tunnel?" He added, "It's hard to break through that entrenched thinking about rail . . . we hear that whistle and are nostalgic about rail."[50]

Bus rapid transit was just one of many alternatives that had to be considered in the state's analysis and decision-making process. Narrowing down the options and variations would become an arduous, multiyear endeavor that continued to be impaired by limited resources, conflicting goals, interagency conflict, lack of leadership, uncertainty about the alternatives, onerous regulations, and unrealistic expectations.

In 2001, Platt had expected that the state would analyze about thirty different options.[51] Instead, the project team started with a list of 156 alternatives based on input from previous studies, its own preferences, and ideas suggested by the public and participating government agencies.[52] Many of these alternatives had been studied numerous times in the 1980s and 1990s. These included new ferry services, monorail, commuter rail, light rail, congestion pricing, converting the reversible lane on the Tappan Zee Bridge to an HOV lane, and adding new highway lanes,

By July 2003, the agencies reduced the 156 different alternatives to 15 that required further study.[53] They eliminated some ideas that had been ruled out in previous studies, such as requiring corporations along the corridor to charge their employees for parking. Just as the governor's task force had done, the project team dropped the idea of dedicating one of the four lanes on the Tappan Zee Bridge to buses or high-occupancy vehicles, because the bridge was already at capacity during peak periods.[54]

Among the fifteen alternatives making the cut were various combinations of either rehabilitating the Tappan Zee Bridge or replacing it and accommodating bus lanes, monorail, light rail, or a commuter rail line.[55] Although Metro-North and the Thruway Authority were able to agree on fifteen, they would not be able to narrow down the list much further.

Stalemate

In 2003, the agencies started learning more about the cost of building a new bridge and rail line.[56] The consultants provided construction-cost estimates that

Table 7.1
Bridge Replacement and Repair Costs

Alternative	Cost
Repair the Tappan Zee Bridge	$2.0 billion
New eight-lane bridge	$3.3 billion
New eight-lane bridge with bus lanes	$3.9 billion
New eight-lane bridge that could accommodate commuter rail	$4.7 billion

SOURCE: Earthtech, "Level 1 Screening Charette Workshop," materials distributed to charette participants, New York, May 20–23, 2003.

were much higher than those generated by the Thruway Authority and Metro-North in the 1990s and the Governor's I-287 Task Force in 2000. The estimates had not skyrocketed because of inflation, but rather because the consultants conducted much more detailed analysis, included a contingency factor to deal with unexpected costs, and provided more honest numbers. As shown in table 7.1, these projections only included the cost of a new bridge; a new commuter rail line would add billions more.[57]

When the governor's task force recommended replacing the bridge and adding rail, it estimated the project would cost about $4.1 billion. That number had been used repeatedly by the media, legislators, and business leaders. In 2004, when the agencies released their estimates to the public, one number in particular ricocheted around the state: the cost to replace the bridge and add mass transit could exceed $20 billion if all the various transit elements were included.[58]

The high costs exposed an ongoing difference between the counties on both sides of the river. Should new rail services focus on providing Rockland and Orange County residents west of the river with a one-seat ride to Manhattan, or should it serve Westchester workplaces?

For Rockland County and its neighbors to the north and west, the biggest benefit of building a new rail line would be to reduce commuting time to New York City and make those trips more reliable. It would spur both new residential and commercial development. Al Samuels, president of the Rockland Business Association, said, "Rockland must have a one-seat rail link to Manhattan in order to realize its economic future. We must have rail, and that rail must connect Stewart Airport, Rockland, and New York City."[59]

However, just as in the 1990s, Westchester County officials were concerned that transportation improvements might be disruptive and offer few benefits. Instead of I-287 continuing to serve as an important corporate corridor, it could become more of a crossroads for regional travel. Westchester County Executive Spano was not a fan of connecting a new railroad bridge with Metro-North's Hudson Line. It would require taking valuable property along the waterfront that would mar views from historic properties. Spano did not think he needed to oppose it publicly, though. Since the connection alone was estimated to cost $1 billion, he thought it was highly unlikely that the state would spend so much money for a few thousand riders.[60]

Maureen Morgan, now representing the Westchester County Chamber of Commerce, complained that the transportation agencies were focusing solely on New York City's needs. She said, "What began as a study that focused on east/west traffic congestion has become a study mainly concerned about the north/south movement of commuters, a reaffirming of the old radial rail configuration with Manhattan as the only destination in the region."[61] Likewise, Marsha Gordon, the president of the Business Council of Westchester, stated that the east-west rail component should be emphasized, since the purpose of the study was to reduce congestion in the area.[62]

Despite their differences, Rockland County and Westchester County had a cooperative working relationship, in part because they had a common foe. They were both frustrated with the slow pace of the study and the lack of information coming from the Thruway Authority and Metro-North.

From 2003 through early 2005, Metro-North and the Thruway Authority were unable to narrow down their list of fifteen alternatives, even though they did always seem just about ready to announce their short list of four or five options.

In January 2003, the Thruway Authority's Waite said the agencies would identify a short list of five alternatives by the summer of 2003.[63] In the summer, the *Journal News* reported that the short list was expected to be completed the following spring.[64] When spring came around, the project team said it would identify the final four or five options by the summer, and then two months later, the team announced that the public hearing scheduled to discuss the reduced number of alternatives would be pushed back until the fall.[65]

While the Thruway Authority and Metro-North were struggling to whittle down the number of alternatives, a team of Columbia University graduate students selected just one. After working with a group of planners and engineers, they recommended replacing the bridge and implementing bus rapid transit along the corridor. In their view, bus rapid transit was preferable to commuter rail because of its flexibility and ability to accommodate Rockland's low density and rugged topography. They suggested that if demand greatly increased in the future, a rail component should be considered. They also proposed implementing congestion pricing and promoting transit-oriented development and transportation demand management measures.[66]

Metro-North did not want to discuss the students' proposals.[67] Permut had not invested years of his time and millions of dollars of the railroad's resources to develop a plan for a bus system. Instead, the project team used assumptions intentionally designed to make commuter rail a more appealing alternative than buses. In 2004, the Regional Plan Association's Jeff Zupan noticed that the project team had used an overly optimistic frequency of train service and had assumed bus travel times would get much longer even though the commute time by bus hardly had changed at all in the previous thirty-five years.[68] What Zupan did not know was that Permut had been making overly optimistic assumptions about commuter rail along I-287 for over a decade.

A primary stumbling block in narrowing down the list of fifteen alternatives was the agencies' disagreement on a proposal that would create two separate Hudson River crossings—a rail tunnel for Metro-North and a highway bridge for cars and trucks. The consultants thought it would be easier, less intrusive, and less expensive to build a connection to the Hudson Line from a tunnel below the tracks rather than from a bridge above. Given the topography, the engineers had already assumed that much of the rail alignment in Rockland and Westchester Counties would have to be underground anyway.[69]

The Thruway Authority's Chris Waite liked the rail tunnel because separating the two crossings would allow the Thruway Authority to replace its bridge faster. Permut was dead set against that idea, however.[70] His second marching order to Mainiero had been "thou shall not have two crossings." Separating the rail from the bridge would have effectively killed the new rail line, since Permut needed the Thruway Authority to pay for the new crossing.

When Permut studied the rail crossing in the 1990s, he found a rail tunnel to be quite feasible and more popular than a bridge.[71] But now he pushed Mainiero to kill it. He wanted the highway and rail crossings to be "joined at the hip," and he wanted the consultants to write a white paper saying that a rail tunnel was not feasible and to "put it in the file."[72]

On behalf of her boss, Mainiero told the rest of the project team that two separate crossings would be more costly, preclude a rail line across Westchester County, and result in enormous environmental, community, permitting, and construction impacts.[73] She claimed that eliminating the tunnel would guarantee a faster EIS process. Taking a page from Permut, she accused the consultants of producing poor-quality work with misleading information and misstatements.[74]

Waite argued that the extent of the tunnel's impacts should be studied in the EIS and not be left to be speculation. With the two agencies at a stalemate, the project team's attorneys took Waite's side. They said that if someone sued the state, the project team would not have enough evidence to show why the rail tunnel alternative was eliminated.[75]

Through the end of 2004 and the beginning of 2005, the consultants continued working on a report that examined the pros and cons of a separate rail tunnel. In early 2005, Waite stopped the consultants from conducting any more work; he felt Metro-North "wanted the consultants to hammer away at the tunnel until they killed it."[76]

Metro-North president Peter Cannito was unable to solve the stalemate with his Thruway Authority counterparts, John Platt and Michael Fleischer. The relationship between the agency heads was consumed by the need to deal with trivial matters over which their staffs argued. They talked about whether the consultants were objective or biased toward the Thruway Authority, whether the authority's staff was unilaterally giving directives to the consultants, and whether minutes were deliberately misleading Metro-North's position.[77]

NYS DOT to the Rescue

By 2005, Governor Pataki was hearing from multiple sources that the relationship between the two agencies was dysfunctional and slowing down the project.[78] The Rockland and Westchester County executives set up a fifteen-member group to track the state's progress and prompt the state agencies to pick up the pace. This Westchester-Rockland Tappan Zee Futures Task Force was also designed to keep the public informed and generate even more ideas about how to address the transportation problems.[79]

Pataki and his assistant secretary for public authorities, Robert Zerrillo, said they tried to bring the agencies together. Pataki explained how "every agency is a guardian of its own fiefdom" and that he would always try to break the "silo mentality."[80] However, Zerrillo remembered that "they would go back to their silos."[81]

Senior state officials thought NYS DOT could do a better job managing the study than either the Thruway Authority or Metro-North. Zerrillo thought the Thruway Authority had neither the expertise nor the vision that was needed. The Thruway Authority typically relied upon NYS DOT's expertise for its larger projects since NYS DOT had about 2,700 engineers while the authority only had about 60. Zerrillo did not get the sense that the Thruway Authority appreciated the issues related to public transportation and the overall regional transportation network.[82]

NYS DOT's Dennison tried to mediate between the two agencies, but he was unable to find common ground. He realized that one problem was that both agencies were accustomed to controlling their own rights-of-way. The Thruway Authority owned its own toll road with limited entrances and exits. Metro-North operated on rail lines that were only used by Amtrak and Metro-North trains. In contrast, NYS DOT always had to cooperate and compromise with local officials since its roads connected directly with tens of thousands of residential, commercial, and public properties.

Carrie Laney, the governor's deputy secretary for transportation, pushed to have NYS DOT take a leading role for two reasons. First, the state had just lost a lawsuit relating to a new bridge because a judge deemed its EIS to be deficient; she wanted to make sure that Tappan Zee EIS did not meet the same fate. Second, she wanted NYS DOT officials to have a seat at the table because "they were not vested in connecting five rail lines and it wasn't their bridge."[83]

Everyone seemed to think that bringing in NYS DOT was a great idea. Permut lobbied to have NYS DOT act as a neutral arbiter and take the contracts away from the Thruway Authority.[84] The Thruway Authority's DePrima said of NYS DOT, "They are huge agency with lots of people to work on it, while the Thruway Authority has a small planning group. DOT is much stronger equipped to manage a project with the community and the MTA."[85] Fleischer, the Thruway Authority executive director, also welcomed NYS DOT. He thought that in order to obtain public support and funding they needed more players to be involved.

Ultimately, former MTA chairman Conway had to convince Pataki to bring in NYS DOT. Conway remembers that Pataki just didn't understand why the two agencies were not getting along.[86] In December 2004, the governor's office decided that NYS DOT would join the team.[87] Decision making would now be a three-way effort.

Dennison told the *Journal News* in February 2005, "We don't suffer from the single focus of the other agencies. The DOT's concern is the whole trip. It goes beyond the highway. It goes beyond the train. There's been significant tension in the project and it hasn't moved as fast as people wanted."[88] One member of the project team quipped, "The governor in his infinite wisdom thought that if two agencies couldn't get along, then a third would help the process."[89]

Tensions between NYS DOT and the Thruway Authority surfaced quickly. Waite said, "The Thruway can't just turn over the study to the DOT. Laws would have to be changed. The board is answerable to our bondholders. We can't just turn over our fiduciary responsibilities." Dennison sarcastically responded that his agency didn't intend to do anything other than manage the project. "I don't know about the bondholder situation but I do know that we built the Thruway and the Tappan Zee Bridge," he said.[90]

NYS DOT Commissioner Joseph Boardman brought in his director of planning and strategy, Tim Gilchrist, to represent the agency on the project's executive steering committee (which also included the heads of the Thruway Authority and Metro-North). Gilchrist was well respected for his technical skills, political savvy, and toughness. To manage the project team's day-to-day efforts, Gilchrist tapped Michael Anderson, a NYS DOT engineer who had been managing the rebuilding of the highway next to the World Trade Center.

Thanks to NYS DOT's intervention, in September 2005, more than five years after the governor's task force issued its final recommendations, the state released the short list of six alternatives that would be evaluated further in the EIS.[91]

With Gilchrist pushing both Metro-North and the Thruway Authority along, the three agencies agreed to abandon the rail tunnel idea. The project team released a ninety-two-page technical report that claimed train service on a bridge was preferable based on costs, construction impacts, construction risks, environmental impacts, property displacements, and security.[92] Although Metro-North did identify valid flaws with a rail tunnel (such as ventilation issues), one consultant said Permut killed the rail tunnel option by making it impossibly expensive.[93]

Table 7.2
Short List of Alternatives

New bridge or rehabilitate bridge	Corridor-wide transit	Cost estimate (in 2004 billion dollars, average of range)
Maintain in existing condition	None	$0.6
Rehabilitate	Low-cost transit improvements	$2.3
New bridge with high-occupancy toll lanes	Bus rapid transit across I-287	$5.8
New bridge	Commuter rail to Hudson Line and bus rapid transit across Westchester	$10.3
New bridge	Commuter rail from Rockland to Hudson Line and light rail line across Westchester	$11.3
New bridge with high-occupancy toll lanes	Commuter rail across I-287	$13.0

SOURCE: NYS Thruway Authority, MTA Metro-North, and NYS DOT, "New York State Thruway Authority, MTA Metro-North Railroad, and New York State Department of Transportation Recommend Six Alternatives for Further Study in Tappan Zee Bridge/I-287 Environmental Review," press release, September 29, 2005.

As shown in table 7.2, two of the surviving alternatives involved repairing the existing bridge. The other four included a combination of commuter rail, light rail, and bus rapid transit on a new bridge. The alternative with a commuter rail line across the corridor cost more than twice as much as the bus alternative.[94]

Two of the bridge alternatives also included high-occupancy toll lanes, a relatively new and promising transportation strategy at the time. Lanes of his type are designed to accommodate vehicles carrying more than one person, like an HOV lane, as well as drivers without any passengers who are willing to pay a higher toll. The Thruway Authority liked them more than HOV lanes since they would be better utilized and bring in additional revenue.

In 2006, the project team worked on details for each of the alternatives. The bridge designers recommended that two new ninety-six-foot-wide bridges be built to replace the existing ninety-foot-wide span. The team also started studying the environmental impacts of a new bridge, such as its effects on the migration of Hudson River fish and the Lower Hudson Valley ecosystem.

NYS DOT's Anderson, the new leader of the project team, continued emphasizing the importance of transit to any solution. He said a new bridge with mass transit was needed to reduce travel times, provide improved access to employers in the region, and enhance safety.[95] In 2006, the state started providing more details about the bus rapid transit option, including identifying locations for bus lanes, six Rockland bus stations, seventeen Westchester bus stations, and new park-and-ride lots.[96]

The study finally was moving along. Although some people were disappointed that the highway tunnel alternative was eliminated, the general public seemed pleased with the choices. Many members of the press and the public were impressed by the project team's presentation materials and analysis, but time would prove the agencies unable to deliver on the public's high expectations.

Ignoring Funding Challenges

The agencies continued to talk about choosing between various multibillion-dollar projects without discussing how they could pay for them. Not only were these discussions not held in public, state officials were not even talking seriously about it among themselves. The state did not even try to answer basic questions such as:

- How much would tolls have to go up?
- Would new tax revenues be needed?
- How would the new bridge fit in with the region's other infrastructure needs?
- How much federal money was expected to be available?

The Thruway Authority had neither the resources on hand nor the ability to borrow enough money to pay for a new bridge and rail line. Even if the tolls for regular commuters were raised from two dollars to fifteen dollars, the Thruway Authority could issue only about $2 billion in bonds to pay for a new bridge.[97] The state had more urgent needs for the ongoing federal highway funds it received, and the federal government was unlikely to provide much if any funding for the commuter rail alternative. Furthermore, neither NYS DOT nor the Thruway Authority had set aside any money in their multibillion-dollar capital programs for replacing the bridge and establishing new transit services.[98]

The MTA and Metro-North could not afford the project either. During Pataki's last year in office much of the New York City subway's infrastructure, such as pumping facilities to take out floodwater and fan plants to remove smoke, was long overdue for replacement. While the Pataki administration studied spending billions of dollars on a commuter rail line for a few thousand people in Rockland and Orange Counties, the passengers at hundreds of decrepit New York City subway stations would have appreciated the MTA's funding a $60 million station renovation project.

Metro-North and the Thruway Authority had different perspectives about who would pay for the project. Initially, the Thruway Authority assumed that Metro-North could obtain billions of dollars, while Metro-North thought the Thruway Authority could borrow enough money to help pay for the rail portion. When the two agencies saw the rising cost estimates and realized each other's financial limitations, it placed an even greater strain on the relationship and their ability to move the project along.

The Thruway Authority's planning director, Mark Herbst, realized that a new commuter rail line would perform poorly on the cost-effectiveness criteria that the Federal Transit Administration used to determine its funding priorities. When he raised the issue, Metro-North did not want to talk about it.[99] Likewise, the Thruway Authority did not want to discuss raising tolls. In fact, Janet Mainiero referred to the tolling issue as the "T word."[100] During my interviews, I realized that Metro-North and the Thruway Authority had false hopes about each other's ability to provide funding because they had never seriously discussed the issue with each other.

Metro-North's staff was not even talking about how to pay for the project with members of its own board of directors. Ernie Salerno, an MTA board member from Rockland County, was a big supporter of building the new commuter rail line. He told me it would bring in "more people, more development, and more buildings" and that it would lead to lower taxes. He said, "The Hudson Valley would just pop."[101]

Salerno thought the commuter rail idea was very practical, and he was unaware of any fighting between the agencies. He said the board members received limited information, even though the board was approving money for "one study after another." When Salerno did ask senior railroad officials where the state was going to get $12 billion to $14 billion, he was told with "federal funds," and that the money could come from different federal agencies. He did not think that was unreasonable, though, considering how much the federal government was spending at the time on the Iraq war.[102]

Salerno figured anything could happen; the MTA had always come up with money when it was needed. He said the MTA board members never spoke about working with New Jersey on the Access to the Region's Core project, even though it would have provided Rockland and Orange County residents with a one-seat ride to New York's Penn Station.[103]

Governor Pataki was pinning his hopes on a public-private partnership. In fact, building and operating a new Tappan Zee Bridge could have been a profitable business if the state gave a private firm the flexibility to raise tolls. However, a commuter rail line would have needed billions of dollars in public subsidies. One transportation researcher told me, "There is not a prayer that private money could have paid for the rail component. It's delusional." He said the governor "believed in panaceas that don't exist."

Continuing to study a rail project that cost more than $10 billion might not have made much sense from a public policy perspective, but it was in Pataki's best interest. Lou Tomson, a member of the governor's inner circle, believed that Pataki was better off letting the two agencies continue studying the alternatives rather than getting involved and making politically controversial decisions.

Tomson, who served as the governor's deputy secretary of public authorities and later as Thruway Authority chairman, told me, "Finalizing a project meant identifying winners and losers. If you don't believe it's a dangerous situation, why have a public fight about something that won't happen."[104]

Tomson thought adding rail was the "righteous thing to do," but that it was unaffordable and not suitable for suburban trips. He was also skeptical that a public-private partnership would be feasible. He believed that the Access to the Region's Core tunnel project connecting New Jersey and midtown Manhattan along with a bus lane on the Tappan Zee Bridge made much more sense.[105]

Pataki's motivations and behavior were indicative of a genial governor and a staff that avoided problems. In his autobiography, Pataki described himself as having a nonconfrontational nature.[106] He was not interested in getting too involved with the I-287 details and he certainly did not want people to think he was involved. A *New York Times* reporter wrote that Pataki "can seem so detached as to be not in full control of his own administration," and *Long Island Business News* called him "the most detached and uninvolved governor in the modern history of New York."[107]

Just as Pataki hid behind the task force during its proceedings, he hid behind his transportation agencies while they were evaluating alternatives. Of the dozens of people I interviewed, only members of the governor's inner circle realized how much he had cared about replacing the bridge and building a new rail line. People who had been working on the project for many years did not even know that Pataki was the impetus behind the task force's decision.

I asked Charles Lattuca, who worked in Governor Pataki's office before managing the study at the Thruway Authority, about Pataki's role during the feud between the authority and Metro-North. He said, "Matters between agencies didn't rise to the governor. Pataki didn't get involved with that kind of stuff. He had people running the MTA and the Thruway Authority, and he expected those people to do their jobs."[108]

Pataki's office had a penchant for avoiding tough issues and was known for having an attitude of "don't bring me problems, but rather solutions."[109] One of Tomson's signature lines when he was deputy secretary was "Where does this go on my who-gives-a-shit meter?"[110] Except for a few moments, such as when the county executives raised a ruckus about the slow pace of the study, the I-287 project did not rate very highly on Tomson's scale.

By 2006, Pataki's focus was no longer on the Tappan Zee Bridge. He was considering a run for presidency and he wanted to accomplish as much of the rebuilding at the World Trade Center site as he could. Around that time, the media began to turn its attention to Governor Pataki's likely successor, Attorney General Eliot Spitzer.

8

Eliot Spitzer Doesn't Have Enough Steam (2007–2008)

"Who wants to be governor in the middle eight years of a project?" Lou Tomson asked.[1] The former Thruway Authority chairman and first deputy secretary to Governor George Pataki explained why there was little political benefit for Governor Eliot Spitzer to get more involved in the I-287/Tappan Zee Bridge planning process. Spitzer would take the flak for controversial decisions, but Pataki would be considered the visionary and a future governor would bask in the opening day ceremonies.

Tomson realized it would be easier to keep studying the I-287 problem rather than making a decision. He said government policies are "like ball bearings in a tube. It's hard to get them into a tube, but they run forever once you do."[2]

Spitzer's term as governor began in January 2007 and abruptly ended fourteen months later. After learning about the I-287 alternatives and questioning various assumptions, he understood the project's physical constraints and its financial challenges. If he had been in office longer, his aggressive style and his familiarity with the project's details might have changed its ultimate direction. Although he did not champion the project, he did take important steps to move it along.

Eliot Spitzer and His New Team

Like his predecessor, Spitzer recognized the importance of expanding the New York metropolitan area's rail network. He had understood how transportation could spur population and economic growth ever since he had written a paper about the history of New York's subway for a Princeton University urban economics class.[3] As a gubernatorial candidate in 2006, he argued that the transportation system needed to grow in order for the regional economy to flourish.

However, Eliot Spitzer was very different from his predecessor. While George Pataki shied away from direct conflict, Spitzer relished it. He was more like former governor Thomas E. Dewey, who had been an aggressive prosecutor of organized crime figures. Spitzer was known for his intelligence, energy, and the ruthless way he exposed corruption at many of Wall Street's most prestigious firms.[4] He will also long be remembered for shouting at the Republican State Assembly minority leader, "I'm a fucking steamroller, and I'll roll over you."[5]

During his 2006 gubernatorial campaign, Spitzer raised expectations about New York's ability to improve the transportation system. He said the state needed to maintain its core infrastructure, keep tolls and fares affordable, and significantly add to the transportation network in an expeditious manner.[6] During his campaign, he also repeatedly said that he would not raise taxes.[7] It would be impossible to achieve all these goals.

When I asked Spitzer whether he had understood the challenges and constraints of improving the transportation system before he became governor, he responded, "Candidates don't have full access to information and shouldn't. The information gap is enormous. The demands and expectations of a candidate are different." He added, "Candidates need to appear thoughtful, but they don't have all the data and the full negatives," and he noted that those running for public office are not expected to know all the answers. He explained that "people test you" in campaigns and want to see "how you approach issues."[8]

It was a remarkable answer, revealing why he thought it was appropriate for politicians to raise false expectations and promise things that they did not fully understand and would be unable to deliver.

In May 2006, Spitzer gave a well-publicized campaign speech to a Regional Plan Association conference at Manhattan's Waldorf Astoria Hotel.[9] He said, "If we do not add capacity to our transportation systems, we will put a ceiling on economic and population growth in New York State. We face this risk in large part because political gridlock and a dysfunctional transportation planning process that has produced too many competing transportation priorities."[10]

Spitzer went on to state that "the author Robert Caro's great biography of Robert Moses legitimately criticizes the autocratic approach of New York's past infrastructure czar. But I have often said that another biography of Robert Moses could be written that would be titled *At Least He Got It Built.* And that's what we need today: A real commitment to get things done—to get it built."[11]

The gubernatorial candidate then said, "Outside of New York City, one of our most immediate priorities is to embark on the construction of a new bridge with a strong transit component to replace the Tappan Zee, a structure now well past its useful life. Replacing the Tappan Zee Bridge is not only essential for public safety, it is critical to the future economic growth and environmental health of the Mid-Hudson region."[12]

Spitzer's financing plans to replace the Tappan Zee Bridge were deliberately vague. He said that doing so would "cost at least five billion dollars and perhaps several billion dollars more if we add rail capacity." He explained that the state had to look at a "robust public sector financing model" and alternative financing mechanisms such as public-private partnerships.[13]

His Democratic Party primary opponent, Tom Suozzi, responded by stating that Spitzer laid out ideas "that sound good except he doesn't talk about how he's going to pay for it."[14] On primary day, voters were given a choice between one candidate who raised expectations and the other one who raised doubts. Spitzer received 81 percent of the vote compared to Suozzi's 19 percent.

I talked with Spitzer's running mate, the former lieutenant governor David Paterson, about raising expectations, and he said that doing so was "part of the political process. You campaign in poetry. Politicians pretend to give themselves responsibilities they don't have." He also said the political process "infiltrates the governmental side. You don't have government, just continuous campaigns. It hurts because government can't get people to settle down."[15]

Eliot Spitzer's campaign slogan was "Day One: Everything Changes," and he promised to fix Albany's dysfunctional government. In the November 2006 election, Spitzer won 69 percent of the votes in the general election, with the largest margin of victory ever in a New York gubernatorial race.

Once elected, Spitzer appointed Elliot Sander to lead the Metropolitan Transportation Authority (MTA) as its executive director. A former New York City Department of Transportation (DOT) commissioner, Sander had helped write Spitzer's 2006 speech to the Regional Plan Association. He also happened to have been friends with Metro-North's Howard Permut. The two met in the early 1990s when Sander headed the New York State Department of Transportation (NYS DOT) transit division, at which time Permut "chewed off" Sander's ear about the benefits of a new commuter rail line across the Hudson River.[16] Sander envisioned the MTA's building several new rail lines in New York City and one across a new Tappan Zee Bridge.[17]

Tim Gilchrist, who had been the NYS DOT executive overseeing I-287/Tappan Zee project director Michael Anderson, was tapped to help Spitzer's transition team. He said, "In the early days of the Spitzer administration, we were taking the world; nothing was impossible. It's hard to explain those heady times." He noted that not only did Spitzer support New York City mayor Michael Bloomberg's proposal to implement congestion pricing in Manhattan, but he also considered implementing a region-wide congestion-pricing scheme in the New York metropolitan area, along with removing tolls in upstate New York.[18]

As part of Spitzer's transition effort, members of his transportation team met with numerous interest groups and elected officials to learn about the I-287/

Tappan Zee Bridge project and to identify steps to move it forward. A memo Gilchrist prepared for the signature of the transportation team's cochair, Mary Ann Crotty, who had been Governor Mario Cuomo's transportation advisor in the 1980s, noted the project's importance to the tristate metropolitan area but warned of its challenges.

> The number of agencies involved and the local interest in the project makes undertaking the project extremely complex and has the potential for many delays caused by the difference in goals of the involved agencies and different statutory mandates of them as well. While it represents a major opportunity to progress as a legacy project for New York State government it also has the potential to be the poster child for interagency conflict and the inability of bureaucracies to get projects done in an expeditious and cost-effective [manner]. . . . The two-year-old cost estimates are being revised but show a project that could cost more than $14 billion to New York State transportation agencies that will face tremendous short-falls in all transportation programs in the next two years. There is no clear way to fund the most expensive options.[19]

The memo recommended hiring a single project executive to lead the project team, which had representatives from the transportation agencies and consulting firms evaluating alternatives and preparing plans. Crotty's memo also recommended giving the governor's deputy secretary the power to resolve disputes between the agencies and establishing regular meetings for elected leaders and interest groups whose support would be crucial to obtaining future funding and enacting any potential new legislation. Spitzer subsequently appointed Gilchrist to be his senior advisor and deputy secretary for economic development and infrastructure.

NYS DOT's Robert Dennison, who had tried to bring Metro-North and the Thruway Authority together on plans for the corridor, said that Gilchrist's tough, directive approach toward the agencies was problematic. Dennison believed that the two agencies were used to controlling their own worlds and neither was good at compromising. He told me, "They can wait you out. You can move them inches, but getting to the goal line is different if you're bucking up against their culture."[20]

Spitzer received information about the ongoing study from Gilchrist and the MTA's Sander, as well as NYS DOT commissioner Astrid Glynn. Although Glynn was officially in charge of the project, she did not know the players and institutions as well as Gilchrist and Sander, since she had recently moved from Massachusetts. She also did not have the same day-to-day access to the governor that Gilchrist did.

Spitzer's lieutenant governor, David Paterson, noticed that "with Spitzer there was never one person in charge, it was always a team effort." The governor appointed two people to head the state's economic development agency, and two others to lead the state's political committee. Spitzer's transition team also had numerous committee cochairs. He believed that "good people matter more than structure" and that hierarchy was relatively unimportant. But Paterson thought having two people in charge of something was an invitation to fight, and required more of the governor's time.[21]

Before Spitzer was inaugurated, he promised major personnel changes at the state's transportation agencies.[22] Although he brought in new people, he also kept on many officials from the Pataki administration, including Michael Fleischer as Thruway Authority executive director. Paterson wanted to get rid of more of the Pataki holdovers, but Spitzer thought he could "convert them."[23]

Too Much Public Participation?

In February 2007, six weeks after Spitzer's inauguration, the project team briefed the media on their ongoing progress. The Thruway Authority, Metro-North, and NYS DOT announced they would hold two open houses later in the month to update the public on their recent efforts. The team's consultants prepared a series of fifty-foot-long maps that laid out the features of each of the six alternatives that had been under consideration since 2005, including various combinations of rehabilitating or replacing the bridge and building new bus rapid transit, light rail, and commuter rail lines.

Westchester County Executive Andrew Spano and Rockland County Executive Scott Vanderhoef were ticked off when they heard about the press briefing because they felt that they had not been adequately consulted about the alternatives beforehand. Although Thruway Authority chairman John Buono subsequently apologized for not reaching out to the counties and promised it would not happen again,[24] that was not good enough for Spano and Vanderhoef.

The county executives had previously taken two high-profile steps to ensure they were part of the planning process. In 2001, Spano delayed a vote at the New York Metropolitan Transportation Council until the state set up an Inter-Metropolitan Planning Organization that would be regularly updated and would sign off on various proposals for the bridge and corridor. In 2005, after growing increasingly frustrated at the study's pace, the two county executives set up a fifteen-member Westchester-Rockland Tappan Zee Futures Task Force to work with the state's project team and keep the public informed.

The February 2007 press briefing was another breaking point for the county executives. With a new governor in office, it was an opportune time to express their concerns about the project team's lack of progress. Spano and Vanderhoef asked Spitzer to intercede and assign someone to oversee the project. They called the current planning process confusing, unresponsive, poorly handled, and unreliable. They said it was coming across as "bureaucratic, provincial, and lacking any sense of regional vision."[25]

The county executives were particularly concerned over the lack of imagination in addressing transit alternatives. They told Spitzer, "Strong leadership must be provided by your office that is not vested in the future of any one agency, but rather would look at the large picture to catapult New York State to the forefront of economic development and mobility."[26]

Spitzer took the county executives' concerns seriously. He thought highly of both of them, and the views of his friend and fellow Democrat Andrew Spano were politically important. Spitzer realized the county and state had very different interests, though. The county executives were worried about local issues, while he was concerned with expanding the regional economy.[27]

At the start of the Spitzer administration, the bi-county task force leaders felt they could make a valuable contribution to the process, but the state agencies were shutting them out. After attending the public open houses and reviewing the state's materials, they grew even more concerned. In April 2007, the task force's cochairs told the county executives, "These concerns alarm us and highlight the need for action by the governor immediately," and that "The study was moving away from a conclusion that would best serve the residents and businesses of Rockland and Westchester counties, not to mention New York State and the wider region."[28]

The task force leaders had a long list of concerns. They thought that the project team was not paying enough attention to a number of key regional transportation elements, such as a bus rapid transit system east of White Plains, access to Orange County's Stewart Airport, the Access to the Region's Core project under the Hudson between New Jersey and New York, land use issues, parking, freight issues, and connections with existing bus services. They believed "the six alternatives were at least partly selected to achieve compromise among the state agencies, and not necessarily on their potential benefits to the mobility needs of the public." They wanted the project team to revisit some of the options that had been discarded, including the full-corridor light rail line, the rail connection to Metro-North's Harlem Line, and a new tunnel under the Hudson.[29] Although the counties were concerned about the project team's sluggishness, they were making it exceeding difficult for the state to make any progress at all.

At every community meeting, the I-287/Tappan Zee project director, Michael Anderson, talked about how the state valued and wanted community input and involvement.[30] The project team undertook extensive public participation efforts because federal regulations required it, but they also did so because Pataki and Spitzer had thought it was important.

Just as NYS DOT commissioner Franklin White had sought the Westchester County executive's blessing to build a Cross Westchester Expressway HOV lane in the 1980s, NYS DOT commissioner Glynn recognized the importance of having county executive support in the 2000s. One senior state official said opposition from Rockland and Westchester would have hurt her efforts in

Albany. "They would make it harder to get the project done and they would make a project look bad. Once a project looks bad, it loses friends fast."[31]

The state was trying to accommodate too many interests. In some ways, public participation had become a goal in and of itself. Ensuring that everyone was heard and had an opportunity to participate seemed to have become more important than actually undertaking a project. Over fifty federal, state, and local agencies, including more than twenty municipalities along the corridor, had an interest in the planning efforts.[32] It was impossible to satisfy them all.

The project team held open houses, public hearings, public workshops, and project updates. They set up the Inter-Metropolitan Planning Organization and they worked with the Westchester-Rockland Tappan Zee Futures Task Force. They reached out to environmental and regulatory agencies including the Coast Guard, Army Corps of Engineers, National Oceanographic and Atmospheric Administration, Environmental Protection Agency, and the U.S. Fish and Wildlife Service. They held meetings with municipal representatives throughout the corridor to hear the local perspective on issues related to the megaproject.

To provide an open forum for discussion, the team members set up a stakeholder committee with representatives from educational institutions, emergency-services organizations, hospitals, businesses, and other institutions. They compiled a mailing list of more than four thousand stakeholders, set up a website with thousands of pages, and published newsletters and fact sheets. They briefed the media and elected officials and met with representatives from the Palisades Center mega-mall, the Regional Plan Association, the Tri-State Transportation Campaign, and environmental groups. They even set up two community outreach centers in Westchester and Rockland Counties, staffed six days a week, to provide opportunities for community groups and individuals to obtain study information and provide feedback.

But no matter how extensive the public participation efforts were, it was never enough. There was always a call for the project team to provide more information, conduct further analyses, consider more options, and hold more meetings. Involving so many players made generating a consensus exceedingly difficult.

Skepticism, Suspicions, Conflict, and Confusion

The state was weighed down because its planning process was plagued by skepticism about the value of commuter rail, the community's suspicions about the Thruway Authority's intentions, inconsistent opinions from the civic community, and confusion about complex alternatives.

In my interviews, I encountered few transportation professionals outside of Metro-North who thought building a commuter rail line from Rockland County to Manhattan was an efficient use of public resources.

Neil Trenk, a senior transportation planner in Rockland County's Planning Department, thought the rail alternative was not feasible given the "massive grades on both sides of the river and the need for tunnels and viaducts across the county." He said, "I don't think advocates understood that." Naomi Klein at Westchester County's transportation department thought a new rail line was not practical because it was "expensive to have an extensive feeder system" for scattered origins and destinations. Metro-North's I-287 project manager, Janet Mainiero, remembers that the team's consultants "saw multiple problems with rail, especially commuter rail. The consultants liked bus rapid transit; they were pumped up about it."[33]

Jeff Zupan at the Regional Plan Association thought the project team was overstating the benefits of commuter rail ridership and understating the benefits of bus rapid transit. He believed that rail along the corridor was "a nonstarter" because of its cost, limited ridership, political infeasibility, and the need for miles of tunnel and viaducts to go across the ridges and valleys.[34] He told Anderson that the bus lane leading to the Lincoln Tunnel carried up to thirty thousand New Jersey commuters in an hour, almost matching the peak-hour volumes by train from Westchester to Grand Central or from Long Island to Penn Station.[35]

The Tri-State Transportation Campaign argued that bus rapid transit was preferable to rail because buses could pick up riders near their houses, enter the busway to beat congestion, then leave the busway to drop riders off close to where they work, all without requiring a transfer.[36] Jim Tripp, the Environmental Defense Fund attorney who started Tri-State, said, "The rail plan was ridiculous from day one." He did not even think a full bus rapid transit was needed, but rather incentives for vans and carpoolers.[37]

A highly regarded planner in NYS DOT's Transit Bureau realized that many people were approaching the choice between rail and bus in the wrong way. He said planners needed to consider the market for potential transit services and how that market would best be served. He thought a rail line was "was totally unsustainable" and had become a distraction that undermined looking at feasible transit alternatives. He thought a bus rapid transit system was preferable because it could be enhanced and expanded.

Other leading transportation officials were shaking their heads. Tom Schulze, the former executive director of the New York Metropolitan Transportation Council, said, "Nobody outside of Metro-North thinks it's justified. Despite all the analysis, it is not serving a need." He said, "Nothing in the analysis points to commuter rail." Sandy Hornick, who led New York City's efforts to extend a subway line from Times Square to Manhattan's far West Side, said the rail line was "crazy" for the Tappan Zee corridor, noting that while bus was the logical thing to do, rail was sexier.[38]

A federal highway official asked NYS DOT's Rich Peters why Metro-North wanted to build a rail line over the Hudson River instead of "just making improvements to the existing rail line and using Access to the Region's Core?" Peters explained that "Metro-North doesn't want to contract with

New Jersey Transit anymore." The federal highway official responded, "Oh, now I understand."[39]

Rail advocates had a different type of skepticism when state officials mentioned that the bridge might be built with the capacity to add a rail line at a later date. They knew that even though the George Washington Bridge was designed in the 1920s to accommodate a second level for trains, the Port Authority instead constructed a lower deck for six additional vehicle lanes.[40]

The qualms about the rail line were matched by suspicions about the Thruway Authority. At public forums, community leaders from Rockland County's riverfront villages often mentioned the secretive nature surrounding Governor Dewey's decision to build the Tappan Zee Bridge near Nyack. They were suspicious of the Thruway Authority's intentions, its seriousness about including transit, and the need to replace the bridge. One senior Thruway Authority official felt as though he was on the hot seat. He said the community thought "we owed them something for the way we built the bridge in the first place."[41]

The Thruway Authority's own actions did not help build trust. For years, the authority rebuffed requests from community leaders for an independent examination to assess whether the Tappan Zee Bridge really needed to be replaced. Community activists were frustrated that no independent engineers were ever given access to bridge documents despite repeated requests.[42]

Members of the Tappan Zee Preservation Coalition, a community organization long opposed to any new crossings, felt the Thruway Authority was always coming up with different excuses for building a new bridge. For example, the authority's engineers talked about marine borers eating the causeway's wood foundations, but the group discovered that no borers were actually ever seen near the bridge.[43]

One thing that really annoyed the coalition was the Thruway Authority's assertion that the bridge was designed to last for fifty years.[44] The claim was widely reported in the media, including the *Journal News* and the *New York Times.* The coalition asked for proof, but the Thruway Authority was not able to find a single document to support the claim. NYS DOT's Anderson later told me the assertion was an "old wives' tale."[45]

Although project managers stopped referring to a fifty-year design life, the idea stuck and continued to appear in numerous media stories. By the time the bridge's fiftieth birthday rolled around, the Tappan Zee Bridge had become a symbol of the state's inability to care for its infrastructure.[46]

The environmental community did not provide a consistent and uniform opinion about the alternatives that would have helped create a consensus on a practical plan. The Federated Conservationists of Westchester County, the coalition of Westchester environmental groups, supported a commuter railroad. They claimed that connecting the existing railroad lines would be the best way to neutralize suburban sprawl and reduce air pollution. Board member and former president Maureen Morgan, said, "It'll be just money down the rat hole if they choose anything else."[47]

Other environmental groups had different opinions. The Rockland County Conservation Association was not convinced that commuter rail was the best choice, and Riverkeeper, concerned about suburban sprawl, favored rehabilitating the existing bridge.[48] Meanwhile the Tri-State Transportation Campaign supported bus rapid transit across the entire corridor, although sometimes it supported a new commuter rail line as well.

The rhetoric between the groups could get personal. When the Tri-State Transportation Campaign set up a conference to discuss bus rapid transit, Morgan stated that it was "astonishingly arrogant" to focus on only one of the transit options under consideration. She claimed that Tri-State was "trying to shove BRT right down our throats." In its *Mobilizing the Region* newsletter, Tri-State responded that "Ms. Morgan's definition of arrogance seems to apply to any viewpoint not in complete agreement with her own—she and the Business Council of Westchester long ago decided that cross-county rail was the only answer for the corridor, facts, analysis, and the real world be damned."[49]

Meanwhile, Metro-North and the Thruway Authority continued to argue. For example, they could not agree on the location of the rail tracks along the Thruway's right-of-way. Metro-North wanted the tracks in the median because putting them along the side would have required the reconstruction of interchanges and bridges at a cost of approximately $2 billion.[50] Metro-North also wanted its tracks as far as possible from people's homes in order to minimize community opposition. However, the Thruway Authority did not want railroad tracks in the median because that would make it harder for emergency vehicles and tow trucks to turn around.

One NYS DOT commissioner thought the meetings between Metro-North's president and the Thruway Authority's executive director were counterproductive, since the two agency heads tended to bicker about mundane issues such as who was taking too long to provide comments on a report or who was supposed to pay for certain work performed by a consultant.

Carrie Laney, who worked in Governor Pataki's office and later at the Thruway Authority, said, "It hit me at some point that we should come up with one alternative. Trying to bring six alternatives through the EIS [environmental impact statement] process was fundamentally flawed." The team was overwhelmed trying to choose between rail and bus options, identify routes and stops, determine costs and benefits, and identify the environmental impacts for so many alternatives and permutations. Laney added, "The only thing we could agree on was to keep studying the six alternatives."[51]

In April 2007, when the two county executives publicly criticized the state's efforts and called on the project team to revisit many of its decisions, they raised the public's expectations about the alternatives and heightened the public's skepticism about the state's analysis. They wanted the state to revisit alternatives that had previously been eliminated, making it even harder to further narrow down the number of alternatives. The county executives' efforts exacerbated the public's confusion about a complex process.

The *Journal News* editorial writers picked up on the concerns of the county executives and their bi-county task force. They wrote, "The Westchester Rockland Tappan Zee Futures Task Force rightly points out an apparent lack of integration with other transportation concerns in the region. It also correctly calls attention to a whittling of ideas for the corridor that are popular with residents, though possibly not with the transportation agencies involved. This project must be integrated into the region's current and future infrastructure. It must put the needs of the communities it affects first, not the convenience of the agencies involved."[52]

The *New York Times* editorial writers echoed the county executives when they opined, "After years of languishing under the lackadaisical George Pataki, the project needs Mr. Spitzer's focused energy to get pointed in a clear direction. And then to get moving." The *Times* was also very confused about the planning process; its editorial stated:

> We challenge anyone to go to the planners' web site, www.tbzsite.com, and make sense of the proposals armed with anything less than a three-day weekend and an advanced degree in bureaucratic mumbling. After you examine all the options and "sub-options," ponder the merits of scenarios 4a, 4b and 4c, puzzle over the explanations of the "study process" ("Conduct the Scoping Process and Scoping Meetings for the AA [alternatives analysis] and DEIS [draft environmental impact statement], Formulate a Purpose and Need Statement, Identify Project Goals and Objectives, Formulate Screening Criteria, Develop Travel Forecasting Methodologies, Commence Collection of Baseline Data for the DEIS, Conduct Alternatives Analysis and Public Workshops, Issue Alternatives Analysis Report") you may give up hope, convinced that the project has become lost forever in the land of the engineers.[53]

At the time this appeared, the state was about to take two actions that would involve even more voices and create more confusion. First, the project team established working groups to help the team review the alternatives; about ninety people were brought in to discuss environmental, land use, engineering, bridge traffic, and transit issues.[54] Second, the project team announced that it would undertake a "tiering" approach to evaluating the highway and transit options.[55]

A Game of Chutes and Ladders

In April 2007, Deputy Secretary Gilchrist met with Governor Spitzer to discuss the megaproject's status. They talked about hiring Mary Ann Crotty, the cochair of Spitzer's transportation transition team, as a project executive to manage the process, generate consensus, and resolve disputes. Crotty had stature, experience, and skills that Anderson, the current project director, lacked.[56]

Gilchrist told the governor that the agencies had agreed that NYS DOT would be the project lead. No longer would "all things large and small" require unanimous action of the three agencies. NYS DOT would "consult with and report to its sister agencies" but would be empowered to act efficiently.[57]

The agencies had also agreed to establish a joint project office and transfer the consultant contracts from the Thruway Authority to NYS DOT. Under Governor Pataki, NYS DOT was supposed to have taken over the contracts for the consultants, but the Thruway Authority had resisted. This time, with Gilchrist in the governor's office, the shift happened quickly. To remove some of the conflicts over billing, Metro-North and the Thruway Authority agreed to promptly deposit their share of the consultant's costs with the state comptroller. The consultants' costs were not trivial; they had already reached $55 million.[58]

Spitzer subsequently met with NYS DOT commissioner Glynn, Thruway Authority executive director Michael Fleischer, MTA executive director Elliot Sander, Metro-North's Permut, and Gilchrist to discuss a memorandum of understanding between the agencies. When the governor saw that the memorandum indicated that the deputy secretary would make decisions in the case of an interagency conflict, Spitzer joked, "Hey, I want to decide."[59]

The three agencies agreed to define Metro-North's role as "the study's primary expert on transit issues and as the party that will ultimately be responsible for implementing (directly or with other transit providers) the transit element."[60] It gave Permut yet another opportunity to keep his railroad dream alive.

The agencies expected to recommend one transit alternative internally by Labor Day, obtain public agreement by the end of 2007, and publish a draft EIS by February 2009. The study quickly fell behind schedule, however, because the governor's office could not control the planning process.

Federal laws and regulations trumped Spitzer's power, and the authorities had a will of their own. Gilchrist and others who have worked in the governor's office told me they often had trouble getting the large transportation authorities to follow their direction.[61] Authority officials could resist requests from the governor's office by claiming they needed to get approval from their boards. Often, they were even telling the truth about their board members having minds of their own.

In September 2007, the Thruway Authority's board of directors refused to contribute toward the new office space where staff from the three agencies and the consultants would work together. Chairman Buono told the other members that if they failed to approve the funding request, the authority would look like the new "stumbling block" to replacing the bridge. One of the board members, former MTA chairman Virgil Conway, explained that the funding was important because the project "has suffered from people being on their own turf and too far from each other and out of communication with each other." Buono and Conway were unconvincing, though. The majority of the board members thought the 6,100-square-foot space in Tarrytown was too large and expensive.[62]

Although the project team eventually rented new office space, the state never did hire a project executive. After Crotty turned down the governor's offer to serve in that position, Anderson remained the project director, while Gilchrist and Commissioner Glynn took on some of the responsibilities the project executive would have had.

NYS DOT's preferences remained consistent no matter who was sitting in the governor's office. In the 1980s, NYS DOT determined that an HOV lane was the most effective way to solve the corridor's congestion problem. Under Governor Spitzer, senior NYS DOT officials determined that a bus rapid transit system was the most practical alternative. In fact, a bus rapid transit system is not very different than an HOV lane. They both allow drivers with multiple passengers to access their own lanes along certain segments of their trip.

NYS DOT senior officials knew that the state could not afford the commuter rail alternative.[63] They realized that Albany did not have an extra $10 billion or $20 billion lying around, nor could it even afford the principal and interest payments that would be needed to finance it. Although the project team forecasted relatively low ridership levels for the rail alternative, Commissioner Glynn was skeptical the railroad would even achieve those. One senior NYS DOT official said, "No sane person would say yes to paying an extra billion dollars just to lay the foundation for a rail line over the river."[64]

Establishing new bus services was relatively easy. Routes could be incorporated into villages that were uncertain about a rail line, and the routes could connect village centers. Communities could choose whether they wanted downtown bus stations or bus stations with parking lots on their town's outskirts.

NYS DOT officials knew that bus ridership would be limited, but they believed it could be expanded over time. They realized many people in the New York region did not understand all the benefits of bus rapid transit because buses were considered to be "déclassé."[65] One of their goals was to make the bus option more appealing.[66]

Permut really wanted the rail line, however, and MTA officials had no interest in disagreeing with him.[67] Permut's friend, MTA executive director Sander, supported rail on the corridor based on Permut's "arguments and his track record."[68]

In turn, Glynn had no interest in fighting with the MTA through the governor's office. As a relative newcomer to the state, she could not compete against the MTA's politically sophisticated executives and the rail line's numerous advocates across the Hudson Valley. One senior NYS DOT official told me that she had decided it was "just not worth the fight."[69] The Thruway Authority's executive director made a similar calculation. He needed political, community, and financial support to replace his bridge. Alienating his fellow agency heads would not help his cause.

The three-decade I-287 planning saga was like a losing game of Chutes and Ladders. The players would move a few rungs up the ladder only to find themselves plummeting back to the bottom of the board. In 2007, the state was about to go down a chute deliberately in the hope of going faster up the ladder.

In 2007, the federal transportation agencies were concerned with various elements of the study. They suggested that the state use a "tiered" approach to the environmental review so that it could focus on the bridge first and the transit component second.[70] Federal highway officials also thought that the document defining the purpose of and need for the project was not specific enough to define the range of alternatives that had to be studied. Moreover, they were "raising a stink" because they felt that the state was spoon-feeding them information and not involving them enough in the planning process.[71]

If the state agencies addressed those complaints and suggestions, they could save time in the long run, but it would require restarting the scoping process they had started in December 2002. This meant that the state would need to provide the public and government agencies another opportunity to identify potential alternatives and comment on the evaluation process.

In the autumn of 2007, Gilchrist updated Spitzer on the project. He explained how the tiered approach would allow the state to start constructing a new bridge at the same time it conducted the multiyear analysis required to secure federal transit funding. In a telling statement about the importance of the county executives' support, Gilchrist told the governor that before making any public announcement of this new approach, the county executives needed to be in agreement.[72]

Gilchrist reported that the cost of replacing the Tappan Zee Bridge, even without transit elements, dwarfed traditional public financing mechanisms. He said the state would have to consider higher tolls, fees, and regional taxes, and he noted that although there was significant support for the commuter rail line along the entire thirty-mile east-west corridor, a rail line to the eastern end of Westchester could not be justified, based on its high cost and low ridership projections.[73]

Gilchrist instead focused on the two most promising transit alternatives. The first was a bus rapid transit system across the corridor. The second was bus rapid transit along with a new commuter rail line from Rockland County that would connect with Metro-North's Hudson Line in Westchester. This second alternative met the needs of both counties, but as shown in table 8.1, it was almost ten times as expensive.[74]

Figure 8.1 shows the east-west travel market served by the bus rapid transit line and the north-south travel market served by Metro-North.

Spitzer agreed with the recommendation to pursue the tiered approach while continuing to evaluate the costs and benefits of the various alternatives. Now that the agencies needed to restart the scoping process, it opened up another opportunity to raise the public's expectations. One senior Spitzer official said,

Table 8.1
The Cost of Transit Alternatives

Transit alternative	Capital cost of transit component	Annual operating cost	Number of new transit riders
Bus rapid transit	$0.9 billion	$67 million	15,000
Bus rapid transit and commuter rail from Rockland to the Hudson Line	$8.6 billion	$167 million	25,200

SOURCE: Tim Gilchrist, memo to Governor Eliot Spitzer, December 12, 2007, in author's possession.

"The project was seen as a panacea for all problems; a whirlpool of all desires that got funneled."[75]

In January 2008, the state announced that the environmental review would be split into two phases. The first phase would focus on the Tappan Zee Bridge and identify the type of transit and its route. The second phase would address the details of integrating the transit mode into the communities. This new approach would allow the state to complete a draft EIS in June 2009.[76] Anderson told the *Journal News* that "concerns that we're delaying transit, although understandable, are unfounded."[77]

The project team revised the federally required document defining the project's purpose and need so that it clearly indicated that the agencies sought to provide a river crossing that would have structural integrity, meet current design standards, and accommodate transit. In addition, the document emphasized that the project would improve transit mobility and capacity throughout the thirty-mile corridor.

At the public hearings and during the public comment period in 2008, the team received comments from about three hundred agencies, groups, and individuals.[78] Most people were in favor of building a new bridge, although numerous individuals and organizations told the project team to consider more alternatives and conduct more studies.

A group of legislators and local officials who supported new transit services exhibited a classic "not in my backyard" stance. They did not want the state to take any private property along the waterfront in Rockland or even use the Thruway Authority's own right-of-way because that would put the roadway closer to residential properties. They opposed transit stations in eastern Rockland County because parking facilities would destroy the quality of life in the villages. They were also against a commuter railroad because they believed that its noise would negatively impact the educational experience of a middle school in Suffern.[79]

Some people were concerned about the new tiered approach. For example, Tarrytown mayor Drew Fixell opposed phasing the environmental reviews because he feared that once the bridge received its approval, new transit services might be "pushed far off into the future, and perhaps, not really be done at all." He said, "If we really are serious about mass transit, this approach worries me."[80]

Two-thirds of those who commented on the transit options supported commuter rail compared to less than 20 percent who advocated for light rail and bus rapid transit. The president of the Federated Conservationists of Westchester County, Cesare Manfredi, argued, "The bus rapid transit system is essentially a fancier and more complex HOV lane project, a local solution to a regional problem, and as such is doomed to failure on this corridor." He also said, "The only hope for economic survival is the accessibility of a first-class rail system. We almost have it. It just needs to be completed."[81]

State and federal regulatory agencies offered their comments. These agencies sometimes requested more analysis than the regulations required. For example, the director of the state's environmental office in the Lower Hudson Valley told the team that it must consider the impact to environmental resources outside the thirty-mile corridor.[82] But the project team felt that it could not accurately predict the development that would occur as a result of the project, nor did it want to spend time and money trying.

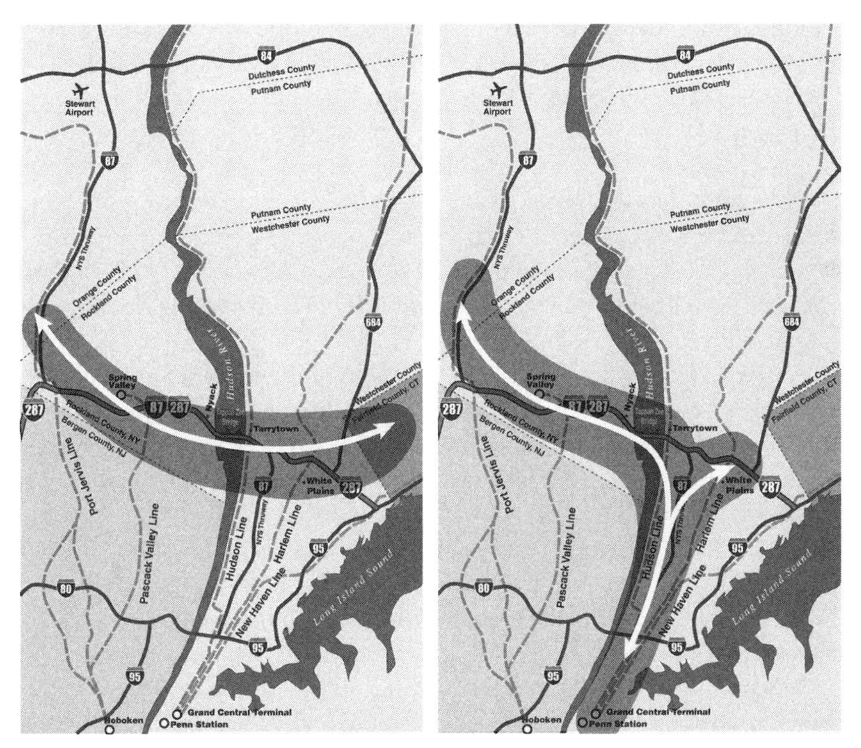

FIGURE 8.1 The east-west travel market would be served by buses (as shown on the left) and the north-south market would be served by trains (as shown on the right).

The End of the Spitzer Administration

Spitzer wanted to end the gridlock and move the I-287/Tappan Zee Bridge project forward. At one point he spent a whole afternoon with Gilchrist learning about the various alternatives and asking tough questions about their feasibility and costs.[83] One senior Spitzer official remembers that "he cared about transportation. He understood that institutional constraints may be bureaucratic but also substantial and that you need consistent willpower to overcome them."[84]

In August 2007, Spitzer told the *Journal News* editorial board, "We're moving the study process forward quickly, trying to build support among the various political leaders who need to be on board to agree on what we should be doing." He went on to say,

> One of the things that happened was there was no one person in charge. There was a nice committee of multiple agencies, but nobody was really taking it by the horns and running with it. Astrid [Commissioner Glynn] is spectacular; she's driving this forward. I've seen the different options. I'm beginning to dig personally into the various cost options and what it means to run bus rapid transit, light rail, traditional commuter rail; what that does in terms of cost, cost structures and transportation flows. So I'm getting an idea, but it's premature to form a judgment. Also, this has got to be a consensus approach. If we don't have the leadership of both Westchester and Rockland on board, it's not going to happen.[85]

He actually had two people (Gilchrist and Glynn) in charge, not one. And he wanted both county executives to support a plan, which was a problem since Westchester's Spano was starting to oppose the widening of any roads in Westchester, even for buses, while Rockland's Vanderhoef was pushing for the costly rail alternative.

Spitzer was not tied to the recommendations of Pataki's task force, but years of study had raised expectations that were difficult for him to lower. He did not think commuter rail was a financially viable option, but he figured that it made sense to continue studying it because it was better for the analysis to kill it than for him to take the blame.[86]

Although the governor heard the engineers' arguments about the bridge's condition, he was not convinced that it really needed to be replaced. He remembers, "One of the challenging premises was that nothing could be done" to rehabilitate it cost effectively.[87]

When I interviewed Spitzer, he brought up the challenges of the "grades and hills" associated with building a commuter rail in Rockland County and connecting it to the railroad's existing Hudson Line. He said, "I don't know technical matters, but I know enough to ask [good] questions." Spitzer also understood the important questions that needed to be asked about the potential for a public-private partnership. He wanted to know, "Where are the savings coming from, who would make money, would there be higher fares and tolls?" It was all a question of economics, he said: "Who is doing well and why?"[88]

Spitzer did not get an opportunity to finalize plans for the corridor. On March 10, 2008, the *New York Times* reported that he had been caught on a federal wiretap arranging to meet with a prostitute. Two days later, the governor announced his resignation. A couple of months after Spitzer resigned, Alexander Saunders, the tunnel advocate, said, "If the Tappan Zee study is any indication of the United States' ability to make rational, innovative long-ranging plans, the twenty-first century is definitely not the American century."[89]

9

David Paterson

The Overwhelmed Governor (2008–2010)

In the five days between Governor Eliot Spitzer's resignation announcement and Lieutenant Governor David Paterson's swearing-in ceremony, the subprime mortgage crisis hit New York. While Paterson was preparing to take office, the market value of the investment bank Bear Stearns plummeted from $3.5 billion to $236 million. A few months later, another investment bank, Lehman Brothers, filed for the largest bankruptcy in U.S. history.[1] At his first State of the State address, Paterson said "My fellow New Yorkers: let me come straight to the point—the state of our state is perilous. New York faces an historic economic challenge, the gravest in nearly a century." He said "the pillars of Wall Street have crumbled," "the global economy is reeling," and "trillions of dollars of wealth have vanished."[2]

Governor Paterson characterized his office as chaotic and undisciplined, constantly moving from crisis to crisis. Some of the crises, such as plummeting tax revenues, were beyond his control, but others were self-inflicted. Paterson told me "I am hugely disappointed in my management style as governor. It's like I had amnesia. I'm a believer in getting rid of institutional problems." Instead, institutional problems festered under his watch. He referred to "Planet Albany" as a place "where thought defies gravity."[3]

Paterson is legally blind, which posed an enormous challenge. He did not have as much time as his predecessors to learn about program details because he had to devote hours to memorizing speeches. Rather than sending memos with charts and maps, Paterson's aides left him short voice mail messages.[4]

Replacing aging infrastructure and expanding the transit system were never Paterson's priority. He did not understand the I-287 corridor like Governor George Pataki, who had grown up in Westchester and represented both sides of the Tappan Zee Bridge as state senator. Nor did Paterson appreciate the project details like the policy wonk Spitzer. Paterson never met with New York State Department of Transportation (NYS DOT) Commissioner Astrid Glynn to discuss the project, nor did he appoint a new commissioner to replace her after she resigned in 2009. One of his closest advisors said he didn't need to worry about the project because "he figured the environmental impact statement (EIS) would take so long anyway."[5]

But like his two immediate predecessors, Paterson saw the opportunity for new rail service to spur economic growth. He told me it would tie Orange County to New York City. He said "Bridges like the Tappan Zee set us apart. Transportation needs to evolve and we need to think like a megalopolis." In that respect Paterson's perspective was similar to that of former Governor George Pataki and Metropolitan Transportation Authority (MTA) Chairman Virgil Conway. Paterson said "If we are going to build a bridge, we should do it the right way."[6]

But at the same time, Paterson recognized the state's debt problem. He said, "We're continuing to bond in logarithmic proportions." He blamed Pataki for "putting the state on a credit card" with balloon payments. Regarding the decision on whether to include a rail line, he told me "There is probably no such thing as good debt, but I'm a believer in an all-inclusive plan."[7]

The Unrealistic All-Inclusive Plan

In June 2008, the MTA announced that Howard Permut would be the next Metro-North president. One of the project's consultants told me that "whenever there was in-fighting between Metro-North and the Thruway Authority, the Thruway Authority never won. Once Permut was President, Metro-North was really number one."

Staff from the state transportation agencies, along with consultants specializing in planning, engineering, public participation and environmental law, had been working nonstop since 2001. Only a few months after Permut became president, this project team put together a plan acceptable to NYS DOT, Thruway Authority and Metro-North, recommending a bus rapid transit system across Rockland and Westchester Counties, as well as a new commuter rail line from Rockland County to Metro-North's Hudson Line.

They agreed that the Thruway Authority needed to replace the Tappan Zee Bridge. Senior NYS DOT officials came to that conclusion because the cost to maintain the bridge seemed to be increasing dramatically with no end in sight, and replacing the bridge would be less damaging to the environment than decades of ongoing reconstruction. The post-9/11 world also influenced their thinking, as they feared a pound of dynamite placed at one vulnerable spot on the bridge would be catastrophic.[8] Tim Gilchrist, the governor's senior transportation advisor, thought the bridge needed to be replaced because the Thruway Authority had failed to follow through on the capital improvement program recommended during Executive Director John Shafer's tenure in the 1990s.[9]

With the economy tanking, NYS DOT senior officials kept waiting to be told to scale back the project and recommend a modest transit improvement instead of the commuter rail option. They asked the governor's office, "Do you want us to be more ambitious or should we say that we're fiscally constrained?" Paterson's liaison responded, "Go for the big one."[10]

Paterson thought an inclusive and open planning process was more important for such a "huge and important project" than rushing to finalize the EIS.[11] So before making any final decisions, the project team set up meetings with the county executives to be followed by briefings to the bi-county task force, elected officials, and stakeholder groups. Now that the team's members had clear direction from the governor, they expected to issue the draft EIS in the summer of 2009.

The team told the county executives why the best transit choice was a thirty-mile-long bus rapid transit system along with a commuter rail line from Rockland County to the Hudson Line. Together, the two transit services would generate higher ridership and offer faster travel times than bus rapid transit on its own. They felt that it was a practical choice compared to the much more expensive rail line that would have gone all the way across Westchester.[12] It also happened to be far less ambitious than Pataki's initial vision of tying in all the north-south rail lines along with Orange County's Stewart Airport.

After the state's transportation officials met with the county executives, the governor said in a prepared statement, "I am pleased the project team and the county executives have come to a resolution on the best way to move forward."[13] One senior Paterson aide said the governor chose an ambitious alternative because even if the state could not afford the project, at least he would get credit for trying to build it.[14]

Paterson's own personality helps explain his decision. Supporting a recommendation endorsed by the project team and the county executives was the easy option. He remembers that "it was always easier to not make a decision." He also recognized that one of his flaws as governor was that he did not want to say no to people. Since he was young, he said, "I realized that I needed to be around people and I never wanted to alienate them."[15]

In September 2008, the project team issued reports to support its recommendations for replacing the bridge and building the new transit system. Permut claimed, "This region will see enormous growth throughout this century and

Table 9.1
Cost Components of the $16 Billion Project

Project component	Cost
Tappan Zee Bridge replacement	$6.4 billion
Highway improvements	$1.9 billion
Bus rapid transit improvements	$1.0 billion
Commuter rail transit improvements	$6.7 billion
Total	$16.0 billion

SOURCE: NYS DOT, "Proposal for Tappan Zee Bridge & I-287 Corridor Unveiled," press release, September 26, 2008.

needs a transit system that will help it grow in a sensible and environmentally sustainable manner. Our recommendation that the cross-county corridor be served by bus rapid transit and that the *huge* travel market to and from New York City be served by new commuter rail service will provide a transit system for the 21st century."[16]

The total project was estimated to cost $16 billion, as shown in table 9.1. At the briefing for the county executives, the state's transportation officials compared their bus and rail plan with other transit options, based on various qualitative and quantitative criteria. Once again, officials manipulated an analysis to show the commuter rail line in a favorable light.

In the early 1990s, Metro-North's rail-crossing study deliberately overestimated ridership and underestimated costs. In 2000, Pataki's I-287 Task Force skewed its analysis to favor the commuter rail alternative. This time, the project team gave the commuter rail option unreasonably high scores on subjective criteria (such as safety and security), used quantifiable measures that favored commuter rail (for example, aggregate travel-time savings), and ignored data that would have favored buses (including transit operating subsidies).[17] As a result, the alternative that scored the best was the one recommended by the project team—bus rapid transit across the entire corridor and commuter rail over the river.

In both their briefings to the county executives and in their report, the project team highlighted how well commuter rail performed on the "cost per passenger mile" measurement. However, the number of miles was calculated in a way that gave the rail option a distinct advantage. Instead of using the actual distance of the new rail line, the team measured the distance from Rockland to Manhattan so that the commuter rail option would appear to be about four times better than it really was.

Moreover, the report noted that the rail line generated much higher revenues than the bus rapid transit system. This was misleading; the train service would have cost a great deal more to build and operate than buses. In fact, the subsidy for train service was astronomical and no one in the public seemed to have noticed.

A little digging and some arithmetic reveal that taxpayers would have had to provide staggering subsidies to build and operate a commuter rail line. The new I-287 transit system would have cost $911 million per year for fifty years in annualized capital and operating costs. That would have required an ongoing subsidy of $19,848 per commuter per year for the first fifty years. In comparison, the bus rapid transit subsidy would have been $3,704 per commuter per year. Building a rail line with such high subsidies would have had the effect of encouraging even more development on the fringes of the New York region.[18]

Evading the Most Sensitive Issue

The I-287 megaproject could certainly have used more transparency. Although the planning process had an extraordinary level of public participation during the Spitzer and Paterson administrations, many questions were left unanswered. When NYS DOT's project director, Michael Anderson, was asked by the Orangetown supervisor if an underground train station could be built in Nyack, Anderson responded, "It's not in the forecast, but it's not off the table. If it makes sense, especially from a planning and community point of view, we'll take a hard look at it." When the mayor of Airmont asked about moving elevated railroad tracks away from two senior complexes, Anderson replied, "We would be happy to work with you and the community in refining that design." When a Suffern trustee expressed concerns that a potential bus rapid transit stop would increase traffic in his village, Anderson responded, "We haven't decided yet. When we do decide, we'll be working very closely with all the communities to work out those details."[19]

The most evasive answers, though, were left for funding. Under Pataki, Spitzer, and Paterson, the project team continuously put off a public discussion of financing. For example, state officials in 2004 said figuring out how to pay for the project would come in later stages of the study. In December 2005, the team said a finance study identifying potential funding sources would begin "next month." In 2007, Anderson said the team would explore traditional government financing and public-private partnerships.[20]

In November 2008, the state finally released a preliminary financial report that included case studies of a dozen large transportation projects, the federal requirements for financial plans, and an analysis of Thruway Authority bonds. It provided little insight into how the multibillion-dollar project would be funded, however.[21]

One MTA official later said, "We went through excruciating details on grants and loans that no one understood. And everyone on the team knew if combined, it wouldn't be enough. But it's probably better to show the public

Table 9.2
Cost-Effectiveness Index: Comparison of Megaprojects

Project	Cost
Second Avenue Subway	$14.16
Long Island Rail Road East Side Access	$18.43
Access to the Region's Core	$24.35
I-287 bus rapid transit	$18.65
I-287 bus rapid transit with commuter rail to Hudson Line	$107.86

SOURCE: Tim Gilchrist, memo to Governor Eliot Spitzer, December 12, 2007, in author's possession.

that you're working than just go into hiding and have people think that we're making deals in smoke-filled room. It wasn't easy for the finance folks to go to the public, because there's no good answers."[22]

The evasion continued. In 2009, Anderson reported that the financial advisor was expected to come up with a range of potential funding scenarios. In 2010, with details still undisclosed, he said the team was continuing to work with its financial adviser.[23]

Although the Federal Transit Administration has awarded billions of dollars to other new rail projects in the New York metropolitan area, the I-287 rail project would not have qualified for those funds. Gilchrist warned Spitzer how poorly it would fare on the cost-effective evaluation criteria used by the federal agency.[24] Table 9.2 shows the cost-effectiveness index for the region's megaprojects with a lower number indicating a more cost effective project. The Federal Transit Administration gives poor ratings to projects with indices higher than $28.[25]

These numbers never were released to the public. When the Regional Plan Association's Jeff Zupan asked Michael Anderson if the state had calculated this cost-effectiveness index, Anderson responded, "We're not up to that point yet."[26] Zupan then tried to calculate it himself based on the information the team had published about costs, ridership, and travel time savings.

Based on the number of passengers expected to use the new rail line, Zupan calculated that the service would have to shave seventy-five minutes off existing transit trips in order for the project to be considered cost-effective. Since the average trip was less than seventy-five minutes, Zupan concluded that "riders would have to go into a time machine in order for the project to get Federal Transit Administration funds."[27]

Neither Pataki nor Spitzer wanted a discussion of finances in public. As long as Paterson contemplated running for a full term, he also was not interested in releasing that information. Gilchrist said, "We didn't want discussions of money in public. If we talked about various toll options, it wouldn't get done. We only talk about how much money we need if we don't want do the project."[28] Instead of releasing financial information, Gilchrist tried to convince the county executives to agree to a scaled-down project, but he was unable to get Rockland County Executive Vanderhoef to give up his dream of a one-seat train ride to Manhattan's Grand Central Terminal.

In 2009, a commission set up by Governor Paterson considered the potential opportunities for public-private partnership across the entire state.[29] The commission thought replacing the Tappan Zee Bridge could be a prime candidate for a public-private partnership, but it recognized that the transit components required far more funding than was available.

Since the commuter railroad would operate at a significant loss, there was one way to get the private sector to help pay for it.[30] If real estate developers were allowed to build high-rise residential buildings and office towers near Rockland and Orange County rail stations, they could subsidize the railroad's construction and operating costs. This strategy had been successfully used in other parts of the world. But New York's suburban communities did not want dense urban development and state officials never considered forcing these communities to accept it.

The State Loses Control

In June 2009, more than six years later than originally expected, the project team officially concluded the alternatives analysis portion of the study. At that time, it expected to issue the draft EIS, the next step of the process, in the summer of 2010. That date was hopelessly optimistic though, because the state agencies were unable to manage the environmental review process effectively.

Anderson said, "Most people don't realize how you lose control, even within your own agency. It's human nature. It's instinctive to cover everything and to do it again and again." He explained that delays were caused by the constant pressure put on the project team to consider more options and undertake additional analyses.[31]

The transportation agencies were unable to manage several aspects of the review. They had trouble defining the project, responding to numerous requests, and addressing the federal government's environmental requirements.

Although the project team's members recommended bus rapid transit across the corridor and commuter rail to the Hudson Line, they still were considering various options within this recommendation such as HOV lanes, high-occupancy toll lanes, bus lanes, and busways. If the team had been able to define the project more precisely, it would have been easier to analyze the project's impacts on traffic, noise, air quality, water quality, and other resources.

In July 2010, almost one thousand people showed up at two open houses the project team held to help citizens understand the details of the project. There still were many questions left to answer.[32] Should the bridges for vehicles and rail be side-by-side or on top of each other? What was the alignment of the bus rapid transit route and where would the state build three dozen stations? How could trains connect from the rail line to a rail yard? How would the new rail line connect with the Hudson Line? Should a rail line be built sixty feet over the roadway to clear the West Shore railroad's tracks, or should it go under the rail line and the Hackensack River? Should buses be given a priority at traffic signals on the streets of White Plains?

Not only did the team have trouble defining the project, they also had a problem controlling the size of the project, because it was seen by local officials as a cash cow with sufficient resources to help them fund their own pet projects.[33]

One MTA official said that local officials had an attitude of "What's in it for me? If there's no benefit for me, what can I extract?" For example, officials in the Village of Monsey wanted the Thruway Authority to build a new interchange between Interchanges 14 and 15 to alleviate local traffic. South Nyack wanted the Thruway Authority to build a park over the highway to reconnect the two halves of the village divided during the bridge's original construction. And Orangetown officials wanted the Thruway Authority to fix some long-standing drainage issues.[34]

The project team had its own pet projects. Thruway Authority officials wanted to make traffic improvements at certain interchanges. Along steep highway sections in Rockland County it wanted to build "climbing lanes"—a clever euphemism because it would not engender opposition the way that proposing additional highway lanes would.

The definition of a pet project was certainly in the eye of the beholder. Some highway officials considered the commuter rail line to be a pet project that was adding to the cost of replacing the Tappan Zee Bridge.

Meanwhile, the megaproject kept getting bigger. At one point, the team was evaluating the environmental impacts that would result from building a new three-mile Hudson River crossing, reconstructing fourteen and a half miles of roadway, adding nine and a half miles of climbing lanes, rebuilding seven interchanges, building four train stations and park-and-ride facilities, constructing a three-mile elevated rail line and a two-mile rail tunnel, and demolishing the existing Tappan Zee Bridge.[35] At the same time, the team always seemed to be responding to requests from the counties for additional analysis and more information.

When a Rockland County planning official suggested a potential alignment for the rail line, he reminded Anderson that if the state did not consider all reasonable options, federal officials might not approve the EIS or a judge could force the state to conduct additional analysis.[36] Ed Buroughs, Westchester County's deputy planning commissioner, offered insightful and detailed comments to the bi-county task force and the project team about the bus rapid transit system. However, his comments required the project team to undertake a time-consuming analysis.

Designing the optimum bus rapid transit system was not a simple exercise, and even county officials had differing opinions among themselves. Planners working in Westchester County's Department of Public Works and Transportation thought buses could operate effectively with mixed traffic if some bus lanes were established, traffic signals were timed to provide buses with priority, and fares collected on the street before passengers boarded. Buroughs, representing the Westchester County Department of Planning, did not necessarily disagree. However, he pushed for an entirely separate bus right-of-way in order to get the project team to take transit seriously and force the state to undertake a comprehensive study. To show how serious he was about developing a comprehensive plan, he made what he called a "shock and awe impression" by bringing along seven county officials to one meeting with the project team.[37]

Buroughs realized a bus rapid transit system could encourage transit-oriented developments where residents could walk, bike, and take transit to their destinations.[38] Many people I interviewed (such as Governor George Pataki and Rockland County Executive Scott Vanderhoef) envisioned a new rail line encouraging people to shift from driving to trains. But Buroughs understood something they did not—that transportation and land-use components needed to be integrated in order to reduce automobile use.

It was relative easy for the project team to draw bus routes on a map, but Buroughs asked them to do much more. He wanted them to evaluate potential bus station locations based on existing land uses as well as a site's potential for future trips. He also asked them to consider how bicyclists and pedestrians would access the bus rapid transit system and the proposed bikeway/walkway on the new bridge.

Buroughs was not the only one who talked about encouraging residential and commercial development that facilitated transit and walking. After the Regional Plan Association and the Tri-State Transportation Campaign pressured state officials, NYS DOT held workshops on transit-oriented development along I-287. But many towns did not want the densities that would create this type of neighborhood. Some residents at the workshops were scared off by renderings of tall buildings, while others were concerned that the density would make the highway congestion problem even worse. Although the workshops started a very useful conversation about coordinating transportation and land use, they made it even more difficult to generate a consensus about the project.

There was another dilemma as well. Metro-North wanted large parking lots around its stations to attract commuters from a wide geographic area. That, however, was inconsistent with transit-oriented development principles, which encouraged residential and commercial development next to stations, not vast areas of parking spaces.

The state's inability to finalize the project's details made it even harder to meet the onerous environmental regulations. When the Cross Westchester

Expressway and the Tappan Zee Bridge were first built, newspaper columnists chalked up opposition to parochial interests. The public expected that some people would have to sell their property so that the greater community could prosper. The planning process for I-287/Tappan Zee Bridge project revealed a much more sensitive culture. Everything was sacred—fish, trees, otters, homes, quiet lawns, low taxes, and cheap tolls.

Federal regulations required the team to consider both the short- and long-term environmental effects on numerous factors such as traffic, noise, vibration, and air quality, as well as the impacts on homes, businesses, neighborhoods, views, parks, wetlands, streams, and cultural resources.

One of the most important resources was the Hudson River. The era when General Electric could dump over one million tons of PCBs into the Hudson and towns could treat the waterway like an open sewer was far from over. The I-287/Tappan Zee Bridge project did not necessarily have to avoid hurting every crustacean and owl egg, but the project team did have to document the expected impacts on vegetation and animals living in and near the river. Fish, birds, coyotes, and other wildlife along the river had to be counted. The team even brought in experts from Louisiana and Baltimore to address how construction noise would affect the fish.[39] If the project team had only had to worry about the Hudson River they could have finished their review of river impacts faster. But the thirty-mile-long corridor had a dozen other rivers besides the Hudson.

Federal regulations indicate that an EIS for an unusual and complex project should be less than 300 pages long.[40] But the thirty-mile-long megaproject was so complicated that the project team needed 326 pages just to explain the methodology the state would use to evaluate environmental impacts.[41]

The state's sensitivity to taking any private property and causing any environmental harm caused delay after delay. It required the state to undertake detailed design studies and additional environmental analyses. By the time the new studies were completed, regulators found that some of the data in the original studies had become stale, which meant that the team had to conduct even more research.

Where Do We Get the Money?

After Paterson announced that he would not run for a full term, the governor's office told the project team that it was time to lay out the reality of the cost projections.[42] In June 2010, Anderson updated the Westchester-Rockland Tappan Zee Futures Task Force on the project team's progress and provided them with the most recent cost estimates, as shown in table 9.3.

After Anderson's presentation, the project team's financial studies manager, Phil Ferguson, showed the task force the financing options that would

Table 9.3
Cost Components of the $20 Billion Project

Project	Cost
New bridge and related road work	$9.4 billion
Bus rapid transit	$1.2 billion
Commuter rail to Hudson Line	$9.2 billion
Total	$19.8 billion

SOURCE: Ed Buroughs, e-mail to Kevin Plunkett, June 22, 2010, in author's possession.

be displayed at upcoming public open houses. Ferguson went through a comprehensive outline of possibilities. Some of the options were expected, such as higher tolls and federal funding. Some were considered innovative, such as seeking contributions from Connecticut and New Jersey. The third set of options—a list of potential tax increases, such as a special funding district—"sparked a fire" in the room.[43]

The task force's cochair, Marsha Gordon from the Business Council of Westchester County, said that if the team released a list of tax increases as one of seemingly equal sources of funding, the project team would lose whatever local support existed to replace the bridge. Her cochair, Al Samuels from the Rockland Business Association, noted that the recently approved MTA payroll tax enacted a year earlier was still highly controversial. The two chairs told Anderson and Ferguson that if the press were to ask them about the potential tax increases, they would say the proposals were outrageous.[44]

After the task force threw cold water on the state's financing options, state officials decided not to release them to the public. The project team's presentation to the task force was the first time that it discussed financing options with an outside group. It would also be the last.

A few months before their terms expired, Lieutenant Governor Richard Ravitch, a former MTA chairman, told Paterson, "The time to address the Tappan Zee Bridge is imminent."[45] Ravitch explained that Hudson River cleanup programs had been so effective that marine organisms were now able to thrive in the river and eat the wooden pilings on the bridge's causeway. Paterson replied with a big smile, "Well, we'll either have to solve the bridge problem or re-pollute the water." Paterson enjoyed his quip so much that he told Ravitch, "We're going to have to put this in a book one day."[46]

Paterson remembers, "The issue was where do we get the money? We had a public-private partnership task force, but they didn't seem to want to touch this." He and Ravitch talked about the possibility of getting funds from the Port Authority of New York and New Jersey. Paterson figured that if New York and New Jersey worked together, they would be more likely to obtain federal financial assistance.[47]

After finding out about the Port Authority's twenty-five-mile jurisdiction, Paterson said that if a former governor deliberately built a bridge above the line, then maybe he could simply expand the Port Authority's jurisdiction so that it included the Tappan Zee Bridge. He and Ravitch called Governor Chris Christie's office and left a message for him. Paterson and Christie never did talk directly about the issue; instead, their two lieutenant governors discussed the possibility of New Jersey investing money in the Tappan Zee Bridge and getting half the revenues. Paterson said, "It was an internal conversation. We just asked if they were interested."[48]

In an interview with a reporter in early December 2010, Paterson mentioned the possibility of the Port Authority's involvement.[49] The next day, Christie responded at a news conference, "I'm not inclined to extend the Port region further into New York just to bite off a monstrous expense." Christie added, "I can't make this any clearer to New York than this: Stop screwing with us. You're not going to come and pick our pockets. New Jersey is not going to permit it anymore."[50]

Paterson believes that a joke he had made a few weeks after the two lieutenant governors talked tainted Christie's reaction.[51] In a September 2010 appearance on *Saturday Night Live*, Paterson was asked whether it was true that other senses are heightened for the blind. Paterson replied that indeed, it was true, stating, "Just by my sense of smell, I can tell that there are fifteen people in this audience from New Jersey."[52]

In 2010, Paterson's term was ending with the I-287/Tappan Zee Bridge megaproject having essentially the same status as when both Pataki's and Spitzer's terms had ended: false expectations were making it nearly impossible to generate consensus on a feasible transit component and move the initiative forward.

In November 2010, Ravitch issued a report finding that the state lacked the revenues necessary to maintain its transportation system in a state of good repair and had no credible strategy for meeting future needs.[53] The $28 billion five-year MTA capital program was facing a gap of at least $10 billion, and the MTA was having great difficulty finding funds to complete two projects already well under construction—the first phase of the Second Avenue Subway and the Long Island Railroad's connection to Grand Central Terminal.

A few months later, the chairman of the State Senate Transportation Committee said that NYS DOT was in dire need of more funds to maintain its existing infrastructure. He noted that the department was in the final year of a two-year, $7 billion capital plan that was about $3 billion below its own recommended funding levels.[54] The governor's office figured that the Tappan Zee Bridge would need its own multibillion-dollar capital program, since it could not take away funding from either the MTA or NYS DOT.[55]

In 2012, I asked Ravitch about the feasibility of financing a brand-new $19.8 billion project when the state did not have enough money to maintain its existing infrastructure. He said that the bond market would not allow the Thruway

Authority to borrow the amount that would be needed to pay for the new bridge, let alone a new commuter rail line, and that there was "no magic way to pay for the bridge."[56]

Referring to an option on the table at the time, the possibility of unions investing their pension funds to pay for the bridge, he said, "It's bullshit about union dollars, as if the union trustees could take the risk." In reference to other ideas such as public-private partnership, he said, "It's all bullshit."[57]

Then, with a few words, Ravitch helped me understand numerous decisions made by George Pataki, Eliot Spitzer, David Paterson, and their successor, Andrew Cuomo. As Ravitch looked me in the eye, he said, "You have to understand that politicians want to kick the can down the road. If it saves them from imposing a burden, they will do it. They will all do it."[58]

10

Andrew Cuomo Takes Charge in 2011

Long before Andrew Cuomo was elected New York's fifty-sixth governor in November 2010, a consensus for addressing I-287 congestion had emerged: the state should replace the Tappan Zee Bridge and establish a new mass transit system.

After Governor George Pataki's task force recommended replacing the bridge in 2000, many residents living near the span had been skeptical about the need to do so. However, by 2010 there was almost universal agreement among business leaders, residents, and civic advocates in both Westchester and Rockland Counties that a new bridge was needed. One longtime local elected official said the opposition to a new bridge had "melted away." He said it happened slowly, but almost completely.[1]

Local officials and residents realized that the bridge was functionally obsolete and physically deteriorating. Ever-increasing demand had made it extremely sensitive to even minor incidents such as someone changing a flat tire. Since the bridge lacked shoulders and a median, emergency vehicles could not race to the scene of an accident, and disabled vehicles could not be taken off a travel lane.

The project team told county officials about the bridge's vulnerability to terrorists, earthquakes, and high winds.[2] Some local officials, like the Rockland and Westchester County executives, were privy to photos (including the one in figure 10.1) that revealed extensive steel corrosion on the bridge's main span.

The project team explained that over a 150-year period, building a new bridge would be less expensive than maintaining the existing one.[3] (Notably, the team

FIGURE 10.1 Steel corrosion on the Tappan Zee Bridge had created serious maintenance challenges. (*Source:* NYS DOT, Metro-North, NYS Thruway Authority, "Tappan-Zee Bridge/I-287 Corridor Project," Status Briefing to County Executive R. Astorino, March 5, 2010)

did not report that repairing and maintaining the existing bridge would be less costly than replacing the bridge over a shorter time period.)

The consensus for replacing the bridge was matched with overwhelming support for new transit services. Strategies to reduce automobile use were the foundation of the planning process in the 1980s, 1990s, and 2000s, when the public heard about the need for transit over and over again. In 2008, the project team said transit solutions were needed to accommodate future growth and reduce dependence on the automobile. In 2009, the team reported that without transit, the projected growth in traffic could restrict economic growth and diminish the overall quality of daily life. A few weeks before the 2010 gubernatorial election, the team said, "New transit is only way to relieve congestion and improve mobility in the corridor."[4]

But the consensus on the bridge and transit had not translated into a financially feasible plan. When Pataki canceled the I-287 HOV lane in 1997, he said the proposed initiative had divided the community over how to solve its traffic problems.[5] Over the next thirteen years, the state never developed a practical plan that brought the community together. In 2010, Tri-State Transportation Campaign's executive director, Kate Slevin, said, "Building a new bridge across the Hudson River and adding two new mass transit systems [commuter rail and bus rapid transit] is one of those once-in-a-lifetime infrastructure projects—so large in scale, so expensive, so hard to imagine—it's no wonder it leaves so many people scratching their heads."[6]

In Search of an Effective Public Sector Champion

In June 2011, five months after Cuomo was sworn in, Westchester County Executive Robert Astorino gave the keynote speech at a forum called Replacing the Tappan Zee Bridge—New York State's Ultimate Infrastructure Challenge. Astorino, a former airborne traffic reporter, called on Cuomo to make replacing the Tappan Zee Bridge one of the state's top priorities. With the project seemingly stalled again, Astorino said, "We can't wait forever for a perfect solution . . . if we don't make rebuilding a priority, the future will be filled with nothing but more expensive studies, more traffic congestion, more bureaucratic delays, and more growing safety concerns." Astorino claimed, "It is time for the planners and engineers to put their pencils down. This is by no means a criticism of them. But it is a criticism of the process. In the absence of leadership driving toward a unified vision, the planners and engineers have been asked to plan for every eventuality in the hope that eventually some of their work may be used." Astorino argued that "commuter rail trains over the Tappan Zee would be great to have" but bus rapid transit offered "a more realistic solution."[7]

Governors Pataki, Spitzer, and Paterson never took a strong public leadership role for the project because it was not in their self-interest to do so. It seemed impossible to identify a solution that was both feasible and popular. Taking a more active role would have required them to take the unpalatable steps of finding new revenue sources and jettisoning unaffordable project elements.

Politically, it would only make sense for Cuomo to actively champion the project if he could achieve meaningful results before his 2014 reelection campaign and a potential 2016 presidential campaign. It did not seem possible to accomplish much in a single gubernatorial term, however, given the onerous environment regulations and the project's size and complexity.

Cuomo's predecessors had wanted to reduce highway congestion along the I-287 corridor, but there was no effective and practical solution to the peak-period traffic problem. Building more lanes would entice more drivers, transit could not accommodate the suburban lifestyle, and political leaders were not willing to raise tolls high enough to have a noticeable effect on traffic volumes.

Planners refer to these types of problems as *wicked* because they are "messy, devious, and they fight back when you try to deal with them."[8] The public administrator Sharon Benjamin-Bothwell explains that wicked problems are "recurring, insidious triangles of contradictions that resist permanent or sometimes even short-term solutions."[9]

When organizations face wicked problems, like solving peak-period traffic congestion, they often have two coping mechanisms. The first option is to study the problem. The organizational researcher Jeff Conklin finds that studying wicked problems is the same as procrastinating, though, since little can be learned about solving them.[10] The second option is to turn a wicked problem into one that can be solved. The state had been trying the first option since 1980.

Four governors and twenty-one years later, Governor Cuomo was about to try option number two.

If Governor Cuomo wanted to implement transportation improvements along the I-287 corridor, he would have to overcome the public's overly optimistic expectations. Residents, travelers, business leaders, and local officials were expecting a one-seat train ride from Rockland County to Grand Central Terminal, express bus services to Westchester employer sites, and billions of dollars to fund it from the state and federal governments. They were also expecting significant congestion reduction and ongoing involvement in the decision-making process.

In early 2011, the state was still planning on building a $19.8 billion project even though it had no way to pay for it. The New York State Department of Transportation (NYS DOT) and Metropolitan Transportation Authority (MTA) were confronting multibillion-dollar budget shortfalls and the Thruway Authority was facing "an urgent financial need" because political pressure to keep tolls low had led it to issue risky short-term debt.[11]

In addition to overcoming high expectations about transit, funding, congestion relief, and the decision making process, Cuomo was also confronting a complex regulatory process. In May 2011, one of the members of the county executives' task force, the real estate developer Robert Weinberg, said to me, "The project team seems to have enormous resources. They just keep working—for years and years." He added, "It's hard to understand what they're doing and I'm an engineer and a lawyer."[12]

As a matter of fact, the state's transportation agencies were drowning in paper. The project team's consultants had recently submitted a draft version of the environmental impact statement (EIS) to NYS DOT.[13] In July 2011, a copy of this 10,000-page document was stacked on top of the desks and shelves of Metro-North's long-range planning department, as the railroad undertook a painstaking and methodical review.[14] One planner was carefully reviewing the materials and his supervisors were checking his work. The planning department was also coordinating its comments with other departments in the railroad as well as with MTA's legal team.

A similar process was taking place at NYS DOT and the Thruway Authority. When the three agencies were ready, they would sit down together to consolidate their comments. The consultants would then make changes, which the agencies would again have to review and approve. Once everyone had signed off on the document, it would be reviewed by the two federal transportation agencies, as well as by state and federal regulatory agencies. The Federal Transit Administration told state officials it would take its Washington and New York offices at least a year to review the document.

One MTA planner told me that the federal environmental laws need to be revisited. "It takes so freakin' long to do anything." She said, "We need to find common ground between the way China builds railroads and our process." She added that the whole environmental process "is ridiculous. There must be a better and easier way for a thirty-mile corridor."

The governor's office also needed to deal with a dysfunctional relationship between the state's transportation agencies. One of the senior consultants thought that Thruway Authority officials were trying to sabotage NYS DOT's leading role.[15] Meanwhile, NYS DOT officials thought the Thruway Authority still did not understand the need to accommodate transit. At the same time, Metro-North and the Thruway Authority continued to find plenty of issues over which to bicker.[16]

Cuomo's Management Style and Motivation

Certain aspects of Cuomo's personality and management style made him well suited to break through the logjam and get the project moving. He had come into office promising to turn around Albany's notoriously dysfunctional government. His governing approach, both aggressive and controlling, had the potential to transform the I-287 planning process.

During the 2010 gubernatorial campaign, a *New York Times* reporter wrote that Andrew Cuomo "strikes you as someone who follows his instincts and is comfortable, maybe too comfortable, being in charge." He reported that Cuomo had a reputation for being irresistibly charming and ruthless. A few weeks before the election, former governor Eliot Spitzer said, "The problem that Andrew has is that everybody knows that behind the scenes, he is the dirtiest, nastiest political player out there, and that is his reputation from years in Washington. When his father was governor, he was the tough guy. He has brass knuckles and he played hardball."[17]

Former governor Paterson told me that he did not have Cuomo's Machiavellian skills. He called Cuomo "a master at making deals."[18] Cuomo's style was reminiscent of Thomas E. Dewey, a governor who accomplished much more than building a 496-mile-long Thruway. Both Cuomo and Dewey put in place sophisticated public relationship apparatuses that outrivaled their predecessors, and they were both accused of being dictatorial.[19]

Paterson was impressed with Cuomo's ability to enact legislation. As governor, Paterson wanted to pass a marriage equality law for gay New Yorkers, but he felt that advocates were fighting over who would get credit for the bill instead of lobbying the state senators. Paterson also wanted to create a center for higher learning at four state universities, but some of the legislators were worried that it might be unfair to minority students. He explained to me how Governor Cuomo was able to get both the education and gay rights bills passed. According to Paterson, Cuomo told the chair of the state assembly's higher education committee (who was a lesbian) that if she wanted the marriage equality bill, she needed to move on his education legislation, and he got key supporters of his education legislation to support the marriage equality bill.[20]

Under Cuomo, the governor's office was known for its meticulous control of information.[21] *Crain's New York Business* wrote, "Observers describe an intense management style unlike anything New York has ever seen." In dozens of interviews, *Crain's* found that people inside or close to his administration "portrayed an intense, micromanaging style that has kept his agencies almost perfectly in lockstep with his agenda."[22] One senior state official told me that when Spitzer and Paterson hired agency heads, the governor gave them direction and autonomy. He said senior Cuomo administration officials in the governor's office had a different approach; they told potential transportation agency heads at their job interviews, "You don't run the agency, we do."

An important difference between Cuomo's office and some of his predecessors was his willingness to tackle thorny issues. He was often heard reminding his aides: "You have to be relevant. If you don't jump on the problem, you're part of the problem."[23] It was very different than a saying associated with Governor Pataki's office: "Come to me with your solutions, not your problems."[24]

Championing the I-287 project offered Cuomo two benefits. The first was the potential to create thousands of construction jobs.

Compared to his predecessors, Cuomo had a different perspective about transportation improvements. Pataki had wanted to promote long-term economic growth in a more energy-efficient, environmentally friendly manner. He laid out an ambitious transportation vision to restore, expand, and integrate the region's extensive rail network. Both Spitzer and Paterson emphasized the long-term growth opportunities associated with a megaproject. Cuomo, however, focused more on the short-term jobs benefits associated with transportation infrastructure projects. Andrew Cuomo's father, Mario Cuomo, had a similar approach when he was governor in the 1980s and early 1990s.[25]

When Andrew Cuomo announced his candidacy for governor in May 2010, New York was still trying to recover from the worst recession since the Great Depression. With the state's unemployment rate at 8.3 percent, creating jobs was a critical component of his platform.[26] Four days after he was inaugurated, Cuomo announced that the state would establish regional economic councils "to create jobs, jobs, jobs."[27]

One senior state official told me that the impetus behind the governor's involvement with the Tappan Zee Bridge project was its potential to create a large number of construction jobs. Another senior official said the governor was looking for a big jobs project of statewide significance, one that affected both downstate and upstate. The governor said to one of his senior transportation officials, "When I see cranes, that's a sign of a community that's growing."[28]

The second reason for Cuomo to take on the I-287 problem was that it would be a symbol of his ability to get things done.[29] A few months after taking office, the state passed a budget that not only was widely praised as fiscally responsible but also the first in five years to be completed on time. Though Cuomo was also on his way to enacting a gay marriage law, building a new Tappan Zee Bridge would be an even greater accomplishment. Not only could he take credit for ending years of dysfunctional planning, he could also both literally and physically help bridge the wide gap between downstate and upstate New York.

A political reporter in Albany who covers Cuomo told me that the governor's platform and political narrative are "I get shit done. I get it done. I make government work." He said that Cuomo "is setting himself up for a national run, whether or not he runs."[30]

Speed Becomes Paramount

When Cuomo boarded a barge to inspect the Tappan Zee Bridge thirteen days after he won the 2010 gubernatorial race, he was told about the extensive maintenance work that the span required. He asked rhetorically, "Should you continue to repair this bridge, or replace the bridge, and how much does the replacement cost? Could you actually improve transportation in the region with a replacement bridge that could include rail, for example?" He noted, "There is a looming nine-billion-dollar deficit that this state faces, so the answer is not going to be more money, more money, more money." A spokesman for the New York State Thruway Authority foreshadowed the governor's decision that day when he said, "What we may be looking at is just putting in the infrastructure for mass transit, and not actually putting that wing on."[31]

A few months later, the governor's office took charge of the I-287 planning process.[32] In April 2011, Cuomo requested a briefing on the status of the Tappan Zee Bridge. His assistant secretary of transportation, Yomika Bennett, asked the heads of the MTA, NYS DOT, and Thruway Authority to provide her with information about the various options for the bridge's repair, rehabilitation and replacement.

A flurry of activity followed over the next few weeks. The governor's special counsel, Linda Lacewell, set up conference calls and meetings, including those with public finance experts from the Bank of America. Bennett also brought in Cuomo's top economic development official and his senior aide charged with improving the bureaucracy's efficiency. The governor personally met with his top transportation officials to discuss the bridge on May 20 in New York and on August 4 in Albany.[33]

Unlike previous governors, though, Cuomo did not include the county executives in his deliberations. He had a much more insular decision-making process and was not interested in seeking any more community input.

In the previous nine years, the project team had held approximately 425 public meetings.[34] In 2010, when Paterson was governor, the team averaged about six public meetings per month. In a nine-month period in 2011 under Cuomo, however, the team held only two public meetings, both of which had been scheduled before the governor's office clamped down on its public participation efforts.[35] The project team stopped holding advisory working group meetings, stakeholder committee meetings, and public workshops. It also quietly shut down the outreach centers in Tarrytown and Nyack.

In late August 2011, Lieutenant Governor Robert Duffy gave a peek into the governor's thinking when he said, "A new Tappan Zee Bridge would solve aging infrastructure issues and boost the economy." He also said, "Infrastructure creates jobs; investment in infrastructure creates jobs."[36]

Since a new bridge would help advance Cuomo's political ambitions, he needed the project to show real short-term progress, something that had not heretofore been associated with the three-decade planning process. When the state's transportation agencies were studying various options in the 1980s, 1990s, and 2000s, they considered various factors such as mobility, the ability to accommodate growth, community impacts, and cost-effectiveness. But in 2011 one factor became paramount: speed. Cuomo's mandate to NYS DOT and the Thruway Authority was to do everything in their power to start construction in 2012.[37]

The Cuomo administration made a number of decisions that allowed construction of a new Tappan Zee Bridge to begin sooner. It brought in a new consulting team, pursued a design-build procurement process, and eliminated the transit component. It also abandoned the public-private partnership concept because that would have delayed the start of construction by a year.[38]

In August 2011, the special counsel to the governor, Linda Lacewell, brought in a new consulting firm to finish the planning and environmental review process. The Thruway Authority's outside environmental attorney and the consulting team's bridge engineer had recommended the firm AKRF and its senior vice president, Robert Conway.[39] AKRF specialized in conducting environmental reviews, and Conway, like Lacewell's boss, had a great deal of self-confidence and a track record of getting things done.

Although Cuomo wanted construction to start in 2012, that could not happen unless the state changed the way it awarded construction contracts, and for that reason he wanted the Thruway Authority to hire a single firm to design and then build a new Tappan Zee Bridge. This would be faster than the state's traditional method, in which it hired an engineering firm to design a bridge, then awarded a contract to the construction firm that could build the bridge for the lowest cost.

However, the state could not award a single design-build contract because doing so violated existing laws. Cuomo would try to get legislation passed to allow design-build contracts, even though previous governors had been thwarted in their efforts by New York's politically powerful public employee unions, which were afraid that design-build contracts would lead the state to outsource more engineering work.[40]

Cuomo also scaled back the scope of the project. Since Pataki had set up his task force in 1997, the state had studied interchange improvements, bridge options, new Thruway lanes, and various transit alternatives along the thirty-mile corridor. This time, the governor's office decided to focus solely on the three-mile bridge so that it could expedite the environmental review, minimize construction time, and lower costs.

It was obvious that the state and MTA could not afford to build and operate a new east-west rail line, but the decision to eliminate the bus rapid transit system was driven by a different factor. Although the governor's team and numerous newspaper accounts made people think otherwise, eliminating a bus lane from the new Tappan Zee Bridge was not a money issue, but rather one of scheduling.[41]

By eliminating the transit component, the state would no longer need to work with the notoriously slow Federal Transit Administration. By dealing only with the Federal Highway Administration, the state would avoid conflicting interpretations of environmental and planning regulations and would expedite the federal government's review.

One participant in the administration's discussions believed that the governor didn't fully understand the desire for transit. Unlike his predecessors, Cuomo did not talk about the importance of expanding the region's transit system. While Pataki liked trains from an environmental, economic, and aesthetic perspective, Cuomo had a different passion. He grew up a fan of muscle cars, and he continued restoring and driving them as an adult.[42]

Even without a transit component, Cuomo was still trying to build one of the widest, longest, and most expensive bridges in the world. The new structure would be twice as wide as the existing one. The single-span, ninety-foot-wide Tappan Zee Bridge would be replaced with two spans: one eighty-seven feet wide for eastbound vehicles, and the other ninety-six feet wide for those traveling west. The wider span would be built first and would accommodate three lanes of traffic in both directions while the Thruway Authority demolished the existing bridge. After completion of the second span, both spans would have four lanes of traffic; the first span would accommodate pedestrians and bicyclists, as well as westbound traffic.[43]

Before Cuomo took office, the project team determined that each span would have shoulders on both sides.[44] In addition, one lane in each direction would be designated for buses and possibly other high-occupancy vehicles. The bus/HOV lane would be located on the left or inside of the roadway to conform with interstate highway design standards.

For the sake of expediency, the Cuomo administration made an important design change that caused much public consternation and confusion, almost placing the whole project in jeopardy, because state officials never explained their rationale. Cuomo's team decided that the width of the new spans would not change from previous designs. They did not want to designate a bus or HOV lane on the bridge, however, because that would have brought the Federal Transit Administration to the table and added about six months of additional traffic analysis and review.[45] Federal officials would have required the project

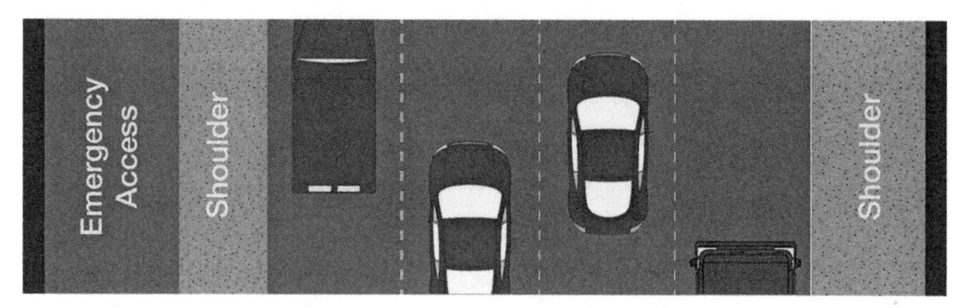

FIGURE 10.2 Rendering of the bridge's southern span with four eastbound lanes along with shoulders and an emergency access lane. (*Source:* Federal Highway Administration, NYS DOT, and Thruway Authority, "Tappan Zee Hudson River Crossing Project," Display Boards for October 25, 2012, and October 27, 2012, public hearings in Westchester and Rockland, http://www.newnybridge.com/documents/meetings/display-boards.pdf.)

team to analyze expected traffic diversions, such as the way buses would travel from the bridge's left lane to the first exit off the bridge.

State officials wanted to claim that the bridge was "designed to accommodate transit" since the two spans would theoretically have enough space in between them to accommodate a bridge with train tracks. However, federal highway officials would not allow the state to exaggerate the bridge's potential. Instead, the federal officials suggested that the planning and environmental documents state: "The bridge is designed not to preclude transit."[46] Those eight words would become the state's mantra for many months to come.

Now that the transit component of the project had been eliminated, the team had to give a name to the lane that had been previously dedicated for buses and high occupancy vehicles. They decided to call it an "emergency access lane," a term that was not part of any standard design.[47] Figure 10.2 shows the 2011 design of the southern span, with four eastbound lanes plus the emergency access lane.

President Obama Provides an Opportunity

Even without the transit component, a bridge replacement project still had to deal with a complex federal regulatory process. In the summer of 2011, President Barack Obama was working on just the thing Andrew Cuomo needed to build his bridge.

On the last day in August, Obama issued a presidential memorandum to federal agencies stating, "At a time when job growth must be a top priority, well-targeted investment in infrastructure can be an engine of job creation and economic growth." Obama asked five of his cabinet secretaries, including Secretary of Transportation Raymond LaHood, each to select up to three high-priority infrastructure projects. The federal government's regulatory agencies would expedite and coordinate their reviews, decisions, consultations, and other actions for these selected projects.[48] The president was looking for projects far along in the environmental planning process that could be expedited, especially those with the best prospects for creating economic growth and jobs.[49]

If LaHood selected the Tappan Zee Bridge as one of his high-priority projects, it would be a godsend for Cuomo, because the state needed approvals or reviews from numerous federal agencies such as the Federal Highway Administration, Advisory Council on Historic Preservation, National Marine Fisheries Service, Department of the Interior, Fish and Wildlife Services, Army Corps of Engineers, Coast Guard, and Environmental Protection Agency. Cuomo's team worked hard to get the Tappan Zee Bridge project to the top of the White House's list.

Howard Glaser, Cuomo's senior policy adviser and director of state operations, lobbied the White House and the United States Department of Transportation (US DOT), explaining to them that the bridge was deteriorating, had an accident rate three times as high as the rest of the New York Thruway system, and was vulnerable to extreme events such as severe storms, ship collisions, and earthquakes. He said the loss of the bridge, or a reduction in its lanes or load limits, would overtax alternate routes and have a catastrophic economic impact on the region.[50]

Glaser told federal officials the project would cost about $5 billion. Tolls on the bridge could pay about $3 billion of that amount, and additional funds would come from labor pension funds and other debt sources, including federal loans. He said the state would cover any shortfall through its general fund or borrowing. Glaser also said the construction project could support well over one hundred thousand jobs.[51]

While Glaser went to Washington, Cuomo personally lobbied Obama's chief of staff, William Daley.[52] New York was the first state to identify and promote a specific project; others were still trying to understand how the streamlining program could help them.[53]

Secretary LaHood took his cues not only from the White House but also from governors, mayors, and local officials. In 2010 and early 2011, LaHood had been deeply disappointed by new governors in New Jersey, the Midwest, and Florida who opposed transportation projects planned for their states. He saw a different attitude in New York regarding the Tappan Zee Bridge.[54]

LaHood recognized that Cuomo, the former secretary of the U.S. Department of Housing and Urban Development, was politically savvy about DC politics. Cuomo's phone calls to Washington sent a signal to LaHood and the White House that Cuomo had made replacing the Tappan Zee Bridge his top infrastructure project and that he would find innovative ways to pay for it.[55]

Cuomo's calls were an important reason why LaHood decided that replacing the Tappan Zee Bridge would be one of US DOT's three high priority infrastructure projects. LaHood told me that when Cuomo took the time to pick up the phone and call senior Obama officials, senior federal officials realized that he was serious about the project. Lahood knew that Cuomo's ongoing efforts would be critical since governors need to provide leadership, vision, seriousness, direction, and financial commitment for megaprojects.[56]

LaHood thought that "all the stars had aligned" for the Tappan Zee replacement project. Not only did it have Cuomo's commitment and federal streamlining, but Congress would pass legislation in June 2012 that would help the project further along.[57] The new law further promoted streamlining and increased the annual allocation of US DOT's loan program for transportation projects from $122 million dollars to $1 billion.

LaHood understood why New York eliminated the transit component. He said Cuomo and his staff recognized that although the administrator of the Federal Highway Administration had successfully streamlined its review process, the Federal Transit Administration's planning and funding process could take twelve years to complete.[58]

The Big Announcements

In September 2011, only a very small group of people knew in advance that the governor's office had changed the scope of the project or requested an expedited review. During Governor Paterson's term, Kate Slevin from the Tri-State Transportation Campaign said the project team was "like some mythical creature living in the depths of the Hudson" making public appearances every six or seven months to dispense information and then disappearing from view.[59] In 2011, the team went down much deeper and stayed there much longer than ever before.

Public notice finally took place over three days in October 2011 via four announcements. On Monday, October 10, the governor's office issued a press release noting that it had asked the federal government to fast-track the Tappan Zee project's environmental review.[60] On Tuesday, the Obama administration announced that fourteen infrastructure projects, including the Tappan Zee Bridge, would be given expedited review.[61] On Wednesday, the federal transit and highway agencies issued a notice that New York would no longer evaluate alternatives and issue an EIS for the thirty-mile corridor.[62] At the same time, the Federal Highway Administration announced that an EIS would be prepared for a replacement bridge.[63]

The flurry of press releases and notices revealed Cuomo's governing style—make decisions behind closed doors, minimize public input, publicize a request only after it has been approved, and ignore the negative implications of decisions.

The Tappan Zee Bridge/I-287 Corridor Project was now called the Tappan Zee Hudson River Crossing Project. It had an estimated price tag of $5.2 billion.[64] The state issued a new schedule in October 2011 that seemed preposterous to anyone familiar with its I-287 track record. The project team was expected to complete its draft EIS within three months and have the federal agencies sign off on the final EIS by August 2012.[65]

The megaproject was no longer trying to achieve Pataki's vision of an integrated rail system. Instead, it was now Andrew Cuomo's most important economic development and political initiative. In interviews and speeches, Cuomo talked about how replacing the span would create jobs and serve as an important symbol of his and the state's ability to get things done.

In November 2011, Cuomo said the state should not approach the Tappan Zee Bridge as a budget issue but instead consider it from a jobs and economic perspective. He claimed, "If we don't approach it that way, I will have to deal with budget deficits for the rest of my term. And the next term will deal with deficits also. We have to turn around the economy. We have to create private sector jobs."[66] A few days later, Cuomo said, "If you replace the bridge you actually stimulate economic development in the region." He said the bridge would generate ninety thousand jobs.[67]

In November 2011, he told a group of supporters, "You know what I say. Build the bridge. Build the bridge. I don't want to hear why we can't. I don't want to hear about the problems. If that was the attitude we'd never be the state. There is nothing we can't do." He said the Tappan Zee Bridge "was a metaphor for the incapacity that we've run into. It's a metaphor for the lack of ability to do big things."[68] According to the *New York Post*, Cuomo said, "I'm going to build the Tappan Zee just to show we can."[69]

Cuomo later said, "I think the Tappan Zee is a good example of a larger problem that we have that is pervasive through this state, I could argue other states across the country also; I could certainly argue Washington. You know, we talk about gridlock—there are different versions of gridlock. We talk about gridlock in terms of legislative gridlock, political gridlock, legislative bodies, they don't act. There's another form of gridlock, which is just the lack of capacity for government, for society through government, to implement big projects."[70]

11

Public Reaction and Cuomo's Campaign (2011–2012)

At first, the public and media praised Governor Andrew Cuomo's October 11, 2011, announcement that the Obama administration would expedite the review process for replacing the Tappan Zee Bridge. The *Journal News* was thrilled that the new bridge "would bring tens of thousands of construction jobs, and it would ease commuting across the Hudson." An editorial proclaimed, "For the sake of our safety and security, our economy and jobs, the time to act is now." *Newsday* editorial writers were delighted because the "project has already been studied to death—to the tune of $83 million—and virtually everyone agrees that refurbishing the rickety old bridge would cost a fortune and still leave the region with an inadequate span."[1]

When local officials and advocacy organizations realized the project's transit component was eliminated, the media changed their tune. On October 16, the *Journal News* remembered that the chief aim of the project had never been to increase vehicle capacity, but rather to build a safer bridge that accommodated mass transit. The editorial argued, "Mass transit needs to remain a part of the solution." An online poll of its readers found a wide majority thought years of planning were "headed down the drain." Likewise, *Newsday* argued that a bridge without transit would not meet the region's needs.[2] The governor now found himself caught between a public clamor for transit and his desire to begin construction as soon as possible.

Confusion, Frustration, and Dismay

Although previous governors and NYS DOT commissioners had emphasized the importance of ongoing county participation and support, Westchester and Rockland County executives Rob Astorino and Scott Vanderhoef never were given a heads-up about Cuomo's announcement. Westchester County learned about the state's plans from the governor's press releases.[3] When County Planning Commissioner Ed Buroughs drove to the project office in Tarrytown to learn more about them, he discovered the office had been shut down.

Astorino was dismayed that the project no longer had a bus rapid transit component. He said, "We cannot build a bridge for the twenty-first century with a bridge that was meant for the twentieth century." Without a mass-transit component, he argued, "you're basically going to open a brand-new bridge after spending billions of dollars and have the same traffic problems you have today."[4] Comparing the governor's plan to obsolete technology, he told state officials "We can't build an eight-track bridge in an iPod world."[5]

The Cuomo administration's tight control of information meant that Westchester County was unable to obtain basic information from the state's transportation agencies. When a reporter asked Astorino in December 2011 how much a bus rapid transit system would add to the project cost, he responded, "We don't know; we can't get even an iota of information of what they're trying to do with the bridge."[6]

Astorino now was concerned that the planning process was moving too fast. He said, "This can't be just a jobs program that doesn't fix the problem. Slow down a little bit. Let's do it the right way. Even if takes an extra year; so we solve a problem for the next fifty years."[7] Ironically, just six months earlier, Astorino was the one who had said, "We can't wait forever for a perfect solution" and that it was time for the planners and engineers to put their pencils down.[8]

No local elected official found himself in a more difficult position than Tarrytown mayor Drew Fixell. On one hand, he was acutely aware of the bridge's flaws and the challenges of trying to repair it. On the other hand, the state was proposing to condemn some private property along the Tarrytown waterfront and build a bridge that would sit closer to Tarrytown homes. To make matters worse, construction of a new bridge would be noisy, disruptive, and cause years of traffic headaches all across his village.

Fixell was suspicious about the arguments used by state officials. He thought the Thruway Authority had for years, made disingenuous arguments about the bridge's unique vulnerability to earthquakes, its inability to replace the deck without replacing the whole bridge, and its purported design life.[9]

He understood why Cuomo jettisoned the project's transit component—replacing a bridge was easy compared to building a thirty-mile transit line. But now he felt there was an element of bait and switch: the promise of new transit services had vanished.[10] Tarrytown was literally at the center of the transit debate. Every day, about one thousand people were driving or taking the bus across the bridge and using Metro-North's Tarrytown train station on the Hudson Line.[11]

Fixell recognized the pros and cons of the various transit options. A new connection to the Hudson Line would have eased traffic near Tarrytown's train station, but it would have marred the views from historic properties along the river. When he learned about the astronomical cost of building an east-west rail line, he tried to convince other members of the Westchester-Rockland Tappan Zee Futures Task Force to support a more affordable transit alternative. However, he was unable to persuade Rockland County Business Association's Al Samuels and representatives of the Federated Conservationists of Westchester County to accept a bus rapid transit alternative, instead.

To accommodate Rockland residents, Fixell suggested building a second Metro-North station in Tarrytown right below the Tappan Zee Bridge. However, the railroad opposed it because it did not want to operate and maintain two train stations so close to each other. At the same time, Fixell opposed the Regional Plan Association's recommendation to build a bus lane directly from the bridge to the existing train station because that would have upset Tarrytown residents along the road's right-of-way.[12] It is no wonder that the state never generated a consensus on a practical transit alternative.

After the state announced its new plans for the bridge in September 2011, Fixell chose his words carefully so as not to offend the governor.[13] Senior Cuomo aides had told him and other mayors that they did not want local officials "throwing darts" at the bridge project. Although Fixell did not criticize the governor, he did publicly express his "profound sense of disappointment" about the lack of transit. He told state officials, "While delay and paralysis serve no positive purpose, there similarly is neither a need nor an apparent justification for rushing forward, as was done in 1952, with a project whose impacts we will live with for the next 100 years or more."[14] Fixell never went beyond what he considered to be acceptable level of critique, and in turn he never felt threatened by the governor's office.[15]

On the other side of the river, Nyack's longtime representatives to the I-287/Tappan Zee Bridge project team expressed their "profound disappointment and dismay at the sudden gutting of the mass transit" option. In a letter to the governor, they wrote, "The recklessness of these actions betrays the long hours of dedicated volunteer work and travel that citizens like us had invested in the project over the previous years. . . . The River Crossing project sacrifices the communities of Rockland and Westchester on the altar of political and economic expediency. This mistake was made once before, in 1955. It must not happen again."[16]

The county executives and local officials were not the only ones expressing their grievances. Transit advocates throughout the New York metropolitan area were exasperated because neither state nor federal transportation officials would say who had decided to remove the transit component and why it was eliminated. The only explanation was found in the state's scoping document, which opaquely included the following phrases:[17]

> "Financing of the crossing alone was considered affordable."
> "It was determined the scope of the project should be limited."
> "The transportation agencies rescinded the previous project."

The Tri-State Transportation Campaign's executive director, Kate Slevin, said that the state erased "ten years of study and consensus in three sentences."[18] She and other transit advocates were also infuriated and confused by the state's cost estimates for incorporating transit on the new bridge. Back in September 2008 and again in May 2009, the project team had estimated that a bus rapid transit alternative would cost only $1 billion.[19] In early 2012, the Cuomo administration started using a new estimate of $5 billion.[20] This new figure was deliberately based on an elaborate and overly expensive configuration.

Westchester County Executive Astorino told a reporter that when calculated per mile, the state's new cost estimates were ten times more expensive than the average bus rapid transit system in the nation. Tri-State Transportation Campaign's Veronica Vanterpool said, "You don't need to dig a tunnel to paint a bus lane" and the Regional Plan Association's Jeff Zupan noted that even though there might not be enough funding for a full-scale bus transit program, the state could still implement some related improvements such as building a bus ramp from the new bridge to Metro-North's Tarrytown train station.[21]

Less than three months after Cuomo's announcement, he was starting to lose the public relations battle because he did not want to slow down the project to accommodate transit. Federal highway officials told state officials that if the Thruway Authority wanted to add a transit component, the state would need to revise its scoping documents and analyze changes to traffic patterns.[22] Even if the state just wanted to designate the emergency access lanes as bus lanes, the environmental review process would take about six months more than had been anticipated.[23]

In January 2012, with the state budget director saying that the governor was "hell bent" on building a new bridge, Cuomo said, "It can't take three years to put a shovel in the ground."[24] It is understandable why the governor would want to finalize the planning process and begin construction in less than three years. After all, reelection day was only two years and ten months away.

To expedite the planning process and limit public criticism, the Cuomo administration conducted only the minimum number of legally required public meetings. Federal regulations required the state to hold two sets of hearings; one kicked off the scoping process in October 2011 and the other solicited comments after the draft environmental impact statement (EIS) was released in January 2012. From October 2011 until July 2012, when the torrent of criticism threatened to stall the project, those were the only public meetings held.

Minimizing the number of meetings certainly was not because of a lack of interest. Some meetings had standing-room-only crowds with more than one thousand attendees.[25]

After Cuomo's October 2011 announcement, the project team invited several hundred members of its stakeholder and advisory groups to a briefing. But when the governor's office found out about the invitations, they told the team to cancel the meeting and disinvite the attendees.[26] Tri-State's Kate Slevin said, "They're making a farce of the public process. All the decisions have already been made behind closed doors." Rockland County's Vanderhoef said, "Uninviting stakeholders speaks volumes; the former process had perhaps too much public involvement, but this one has too little."[27] One of Vanderhoef's aides told me, "It really sticks with you when they announce a meeting and then cancel it."[28]

After the October 2011 announcement, the state also took down the old study's website with its thousands of pages of materials. Slevin said she could not "recall a single example of this kind of wholesale document scrubbing."[29] The state subsequently restored the documents.

Tri-State and Construction Industry Face Off Again

In the months after the announcement, the state's transportation agencies faced a familiar adversary, the Tri-State Transportation Campaign. In December 2011, Slevin told the *Journal News*, "The latest plan to replace the Tappan Zee Bridge is a civics lesson in how not to do a transportation project. It is riddled with problems and does not plan for the future."[30]

In 2011 and 2012, Tri-State advocates used many of the same tactics that had helped them kill the HOV lane in the 1990s. They set up a website, drafted op-ed pieces, wrote letters to the editor, issued press releases, held rallies, and drafted resolutions for local municipalities.[31] Before the state hearings on the draft EIS, Tri-State paid for radio and newspaper advertisements. The ads warned radio listeners, "It's now or never to tame traffic on the Tappan Zee . . . Without transit, we'll be stuck in traffic for decades. Tell Albany we need transit on the Tappan Zee."[32] Tri-State's February 2012 advertisement in the *Journal News* is shown in figure 11.1.[33]

Tri-State established a coalition of eleven elected officials and sixteen organizations to demand that bus rapid transit be incorporated into the state's Tappan Zee Bridge plan.[34] Compared to the 1990s, Tri-State found it both easier and harder to get their message out. With the Internet, communications were now instantaneous, but the media had become more diffused, so the Tri-State staff needed to reach out to numerous community news sources including websites and local blogs.[35] Slevin did not need to contact editorial boards, though, because the newspapers were already advocating for transit.

In the 1990s, Tri-State's leaders were far savvier about community organizing than their NYS DOT counterparts. Since then Tri-State's leaders had grown far

FIGURE 11.1 Tri-State Transportation Campaign newspaper advertisement published before public hearings. (*Source:* Advertisement downloaded from http://www.streetsblog.org/wp.content/uploads/2012/02/TPZ.Print.Ad.FINAL.jpg. Permission granted to use advertisement provided by Tri-State Transportation Campaign's Executive Director Veronica Vanterpool.)

more influential, especially in New York City, where they had successfully promoted the creation of new public plazas and bicycle lanes. But Slevin did not understand that the 2012 fight was very different than the 1990s battle. In the 1990s, Governor Pataki was a tepid supporter of the HOV lane, a project he had inherited from a previous administration. In 2012, Tri-State was fighting much more powerful, passionate, sophisticated, and well-prepared forces.

Ross Pepe, the president of the Construction Industry Council of Westchester and Hudson Valley, had been urging the state to build a new Tappan Zee Bridge since 1978. He strongly supported the HOV lane even after nearly every other official had deserted it. When Governor Pataki canceled the HOV lane in 1997, Ross Pepe had vowed never to lose to the Tri-State Transportation Campaign again.[36]

In 2011, he adopted many of Tri-State's strategies to help Governor Cuomo's efforts. He started a new coalition of more than two dozen organizations that wanted to replace the bridge. He said, "The window of opportunity for this all to take place has finally arrived and we need to act on it quickly. We cannot allow it to escape us. It may never come around again."[37]

Pepe gathered support from the transportation industry, businesses, statewide truckers, and labor unions. Like Tri-State, his group set up a website, lobbied local officials, and developed valuable relationships with the media.[38] The new group claimed to represent more than fifteen thousand employers and three hundred thousand employees across the state.[39]

Unlike Tri-State, Pepe coordinated his efforts with senior Cuomo officials. The governor had close ties to both the businesses and the unions in Pepe's coalition. The construction industry had been generous supporters to Cuomo's campaign and its unions had political clout.[40]

The unions' support was seen as critical in close Hudson Valley local elections because they could mobilize votes, manpower, and contributions.[41] Their union's umbrella organization, the Westchester and Putnam Building and Construction Trades Council, represented a slew of union workers, including ironworkers, teamsters, plumbers, bricklayers, carpenters, electrical workers, painters, roofers, and other trades. Although Tri-State's members were passionate, they were no match for the unions whose livelihood depended upon construction work. The Trades Council reportedly had thirty-six affiliated organizations representing more than thirty-five thousand unionized construction workers.[42]

Cultivating a relationship with construction workers was especially important to Cuomo because he had developed a hostile relationship with the state's public-sector unions. While he was promoting policies to benefit those in the construction trades, he was demanding major wage and benefit concessions from government workers. Not only did the construction unions praise Cuomo's policies, they even helped finance a public relations effort to rein in public-sector union benefits.[43]

Working closely with Pepe and the construction unions was another key figure, Al Samuels, the president of the Rockland Business Association. Samuels helped lead forums, press conference, and rallies supporting the governor's effort to rebuild the bridge and enact design-build legislation.[44]

When Spitzer and Paterson were governors, there was no greater advocate for a one-seat rail ride than Al Samuels. He realized the lack of a direct rail line to Manhattan was hampering Rockland's ability to attract both New York City workers and firms. Samuels said, "They're not willing to come on the bus."[45] He tried to broaden support for the rail line by emphasizing its potential to connect with Orange County's Stewart Airport.

Once it became clear to Samuels in 2011 that the state did not have enough money to build a rail line, he became one of Cuomo's most important advocates. Samuels feared the economic repercussions that would occur if the Tappan Zee Bridge was shut down because it was deemed structurally unsound. Year after year, most players in the I-287 saga held consistent opinions about the various alternatives. So when someone as influential as Samuels shifted his stance, it had a noticeable effect.

Samuels's perspective was colored by his experience on June 28, 1983. When he was driving that day along I-95 in Connecticut, the interstate highway's bridge over the Mianus River collapsed, plunging cars and tractor trailers into the water seventy feet below. Samuels remembers, "I walked up to the edge of the roadway and I never wanted to see that again." He applauded Cuomo's efforts to replace the Tappan Zee Bridge with one that was safer, could accommodate more vehicles, and offered shoulders to improve travel reliability.[46]

After Cuomo's announcement, Samuels's goals conflicted with the interests of Rockland's own residents and even other business organizations. He had little to gain from Vanerhoef's idea to improve bus services over the Tappan Zee; he rhetorically asked me, "Why encourage Rockland people to shop in Westchester?" And, when I referred to the Regional Plan Association as a highly regarded business group, he quickly corrected me and said, "RPA is a highly regarded *pro–New York City* business group. What it's in the best interest of New York City is not necessarily what's best for Rockland and my constituents."[47]

Stretching the Truth

While Slevin was battling contractors, construction workers, and business leaders for public support, the governor's team was making its own case. The state's environmental and procurement documents were carefully worded; flawed methodologies and erroneous statements could lead to lawsuits and lengthy delays. The governor's communications team did not have to hue so strictly to the facts, however, because their statements did not have the same legal ramifications.

The Cuomo administration brought back a claim about the bridge's design life that had been debunked years earlier. The Thruway Authority's executive director said that the bridge "was designed to last forty-five to fifty years"; the governor stated that it "was built as a temporary bridge"; and a Thruway Authority spokesman said it had a fifty-year life span "because that was the design criteria in the 1950s, especially for bridges that went over water."[48] Not one of these statements was true, yet it is not surprising that the claim was made. It had been so widely reported that everyone seemed to think it must be true.

The governor's office also exaggerated the number of jobs that would be created. Its October 2011 press release indicated that every $1 billion invested in construction creates or sustains about 30,000 jobs, implying that the $5 billion project would create more than 150,000 jobs.[49] In November, the governor said the project would create 90,000 jobs and anchor development in the Hudson Valley. He also compared replacing the bridge to the building of the Erie Canal and the development of Manhattan.[50] However, the January 2012 EIS revealed that construction would require only about 2,800 workers per year as well as 2,150 indirect and induced jobs in the state.[51] Nevertheless, a month later the Thruway Authority's executive director used inflated numbers once again when he said that the new bridge would stimulate economic development across the Northeast and create more than 45,000 jobs.[52]

Cuomo's team could intimidate its critics. When State Assemblyman Tom Abinanti recommended building one new span rather than two, the Thruway Authority publicly responded, "Abinanti's uninformed and completely unworkable plan is an indication of just how little he understands about Cuomo's plan." The authority's spokesperson said if Abinanti had talked to the experts like other local elected officials, had "he would understand why his plan is so ill-advised."[53]

The governor deliberately offered false hopes to bolster support for the bridge. A group of Westchester residents and elected officials started a movement to turn the existing Tappan Zee Bridge into a three-mile-long park. They envisioned food stands, sculpture gardens, benches, and summer concerts.[54] Numerous newspaper editorials praised the idea of keeping the old structure and avoiding the need to spend $150 million to remove it. The *New York Times* even asked its readers to submit potential designs for the proposed walkway.[55]

In February 2012, Cuomo knew very well that turning the old bridge into a greenway was not feasible. He had been told that such a plan would cost well over a billion dollars, and there was no room for three separate spans anyway.[56] Still, the greenway concept was helping generate support for building a new bridge, so a few days before the draft EIS public hearing, Cuomo stated that turning the current Tappan Zee Bridge into a crossing for pedestrians and bicyclists would offer outstanding views and recreational opportunities for visitors.[57]

At the public hearings, a Westchester town supervisor thanked Cuomo as well as the *New York Times, Newsday*, and *Journal News* for supporting the proposal to create a three-mile-long park.[58] The cochair of the organization spearheading the idea said she was thrilled by the governor's statement.[59] Eight days later, the Thruway Authority released a document to contractors who were interested in designing and constructing the bridge. A Nyack reporter noticed that the document did not discuss the possibility of saving the existing bridge; instead the word "demolition" appeared twenty times.[60]

Still Avoiding the Funding Question

To minimize opposition to a $5 billion megaproject, the Cuomo administration successfully avoided discussions about how to pay for it. Cuomo was the fourth governor to use that strategy since Governor George Pataki first decided to replace the bridge.

When Cuomo first announced his plan to expedite the project, he characterized the state as getting "creative" about financing. He joked that "a really big bake sale" might be a good way to pay for it.[61] His spokesman said that the state was considering "a variety of funding options."[62] In response to funding questions, state officials typically responded that they were was looking at "alternative financing" and were seeking advice from a financial firm.[63]

New York applied for a $2 billion federal infrastructure loan in early 2012. The program was very appealing to Cuomo in part because its interest rates were lower than those that could be obtained from any other source. Even more important, debt repayment could be pushed off until five years after substantial completion of the project.[64] That meant Cuomo could build the bridge, but future governors would inherit the financing costs. News reports about financing noted that the state and the Thruway Authority would borrow money, but they did not delve into the more important question of who was going to pay off the debt.

The governor's transportation officials did say that financing options included federal loans, federal grants, bonding based on tolls, and pension funds. In February 2012, they publicly mentioned potential toll hikes for the first time, revealing that tolls would be consistent with other Hudson River crossings and include deep local discounts. State officials did not provide any details to clarify these statements.[65]

In May 2012, after the federal government turned down New York's $2 billion loan, the governor said, "We're working through a number of financing options and we'll present a number of options for discussion and we'll pick the best one."[66] The Thruway Authority executive director joked before a presentation about the bridge to the independent Citizens Budget Commission. "I'm going to go through these slides fairly quickly," he said, "because I understand there's a lot of questions on how we're going to finance the project that I can't answer—so I will hedge those after the presentation."[67]

Threat from the County Executives

Although New York's governors are far more powerful than its county executives, Astorino and Vanderhoef had one source of leverage. In order for the project to be eligible for a federal loan or grant, the New York Metropolitan Transportation Council needed to add the Tappan Zee Bridge to its list of projects eligible for federal funds. Both county executives sat on the council, and the vote had to be unanimous. In 2001, Astorino's predecessor had held up the planning process until Westchester County was given an important role. A decade later, the county executives' potential veto threatened to slow down the state's efforts again.

With crowds jamming the public hearings, the county executives were under a great deal of pressure.[68] At a February 2012 draft EIS hearing, I asked one of Vanderhoef's senior aides what he thought the county executive was going to do. He said, "You see all those construction workers out there. He's not going to stand up to them. He doesn't want to stop it."[69] Vanderhoef recognized the unions' political power and credited Ross Pepe for enhancing their influence.[70]

Much to the consternation of Cuomo's team, Astorino and Vanderhoef refused to vote before the state issued its final EIS.[71] It was the governor's own fault, though. In the spring of 2012, the state still was not sharing information with the county executives. Astorino and Vanderhoef were learning about the project from the governor's press releases and the media. The county executives were unable to obtain sufficient information about the bridge's design, its potential for transit, financing options, and expected tolls.

Vanderoef's constituents expected him to have answers, and the riverfront communities wanted his help in their efforts to minimize property takings and disruptions during construction.[72] As Astorino told the media, "People in Westchester are asking me, what about the bridge? What's it going to cost? What are the tolls? What's it going to look like? And I go I have no idea, I'm waiting for the same information that you are. So how could I be expected to vote for something without even seeing it? That's the issue."[73]

The county executives were walking a tightrope. Vanderhoef wanted the transportation agencies to designate the emergency access lanes as dedicated bus lanes during peak travel periods.[74] He was worried, though, that he would be labeled an obstructionist and face retaliation by the governor's office if he delayed the project. He was also concerned that state officials would cut back on Rockland County projects and fail to give him public recognition at Rockland County events.[75] Astorino had another factor to consider. His vote could make or break his statewide political ambitions.

The two Republican county executives found themselves in the politically awkward position of promoting a more expensive solution than a Democratic governor. Larry Schwartz, the secretary to the governor, accused Astorino of being "all over the map" on his position. Schwartz claimed that $30 tolls would be needed to pay for the $5 billion bus rapid transit that Astorino supported.[76]

With Vanderhoef and Astorino delaying the New York Metropolitan Transportation Council's vote, and the media relentlessly attacking the project for its lack of transit, Cuomo's team changed its strategy. On the afternoon of June 28, 2012, the governor met with his communications team and Phil Singer from Marathon Strategies.[77] Singer was well regarded for his crisis management and coalition building consulting services.[78] The Cuomo team would now treat the Tappan Zee Bridge like a political campaign. The minimum amount of public participation would be replaced by extensive community outreach and a public relations blitz.

Only a few hours after the meeting, Cuomo's office announced that the state would create exclusive bus lanes on the new bridge during rush hour as a first step to bringing mass transit to the span.[79] This announcement created a problem for the project team, though, since they were just a week away from wrapping up the final EIS.[80]

Federal Highway Administration officials would not let the state claim in the EIS that buses could use the emergency access lanes, because the project team had never analyzed its traffic and safety implications. Cuomo's deputy secretary for transportation, Karen Rae, needed to figure out how to appease the federal highway officials' concerns. After some back and forth, they compromised on language that was somewhat consistent with the governor's announcement. The EIS would indicate that the bridge "could support the ability for express bus services to use the extra width on the bridge during peak hours. This use would have to be appropriately assessed and considered before being implemented."[81] The EIS language was a big step back from the governor's announcement, but it did not generate any public backlash whatsoever.

Cuomo took a page from Governor Thomas E. Dewey's playbook. In 1950, Dewey hired a United Press reporter to help him generate support for the New York State Thruway. In 2012, the governor hired a local TV news anchor, Brian Conybeare, to serve as his liaison with Hudson Valley residents. Conybeare hosted day-to-day meetings with local leaders and worked closely with the governor's top aides. The popular and friendly TV star was a hit with local officials. Just two days after Conybeare started, the state also set up a phone number for residents' questions and complaints, created a new website and a Twitter feed, and formed two community outreach groups.

In July, Cuomo held numerous meetings with his transportation team and two of his top officials, Secretary Larry Schwartz and Director of Operations Howard Glaser.[82] He assigned Schwartz, who was known as the governor's bulldog and enforcer, to rally support for the project in Westchester and Rockland.[83] As a former Westchester deputy county executive, Schwartz knew many of the local elected officials and the ways in which municipalities relied on state aid. In recent memory, only once before had the governor's chief of staff dedicated so much time to a single project, and that was after the terrorist attacks of September 11, 2001.[84]

Cuomo's office set up dozens of community meetings, with Schwartz attending about one-third of them.[85] In the first three weeks of August alone, the governor's team held fourteen meetings with community organizations, chambers of commerce, realtors, residents, and other groups.[86] At one town meeting, Schwartz even promised to visit every home in a neighborhood where residents were concerned about construction impacts.[87] It was an extraordinary gesture from one of the state's most powerful figures.

Cuomo also rolled out a series of measures to appease his critics. He said the Thruway Authority should expand toll discount programs for Westchester and Rockland residents, and he agreed to set up a transit task force to consider corridor-wide transit solutions. To address aesthetic concerns, Cuomo announced that the panel selecting the bridge's design would include architects, river town historians, international design experts, and local officials.

During the week of August 5, the governor issued six separate press releases announcing support from congressmen, state senators, assembly members, mayors, county clerks, county legislators, council members, former Westchester County executives, and town supervisors.[88] Cuomo himself made many calls to get people behind the bridge proposal.[89]

Although Cuomo still had no interest in releasing any financial information to the public, he found himself with no choice once August rolled around. Federal highway officials had been concerned that much higher tolls could adversely affect low-income residents as well as shift traffic from the Tappan Zee Bridge to alternate routes, so they told state officials that they needed to identify the proposed tolls and evaluate the potential traffic impacts.[90]

When the Thruway Authority released a table of expected toll rates, regular commuters learned that the cost of their daily crossings would rise from $3.00 to $8.40. State officials claimed that tolls for a new bridge would be only modestly higher than those for a repaired bridge, and that a bus rapid transit system would require the authority to double its tolls.[91] It was another dubious claim, however, since the Thruway was building two spans rather than just repairing one, and could implement a relatively low-cost bus rapid transit system.

Cuomo's personal involvement helped bring the county executives on board. Vanderhoef was first elected in 1994, but he had little interaction with any of the four governors. Occasionally, he would see them at functions with other people, and sometimes he would talk to their staffs.[92] So, when Vanderhoef got a phone call from Cuomo in July 2012, it was a big deal.

All the state's meetings with communities and property owners gave the county executives cover to support the new bridge. With the state's newfound sensitivity and attention, local residents and officials were getting somewhat reassuring responses to questions that the county executives had not been able to answer.

On August 20, after several tense months on both sides, the county executives voted in favor of adding the Tappan Zee Bridge replacement project to the New York Metropolitan Transportation Council's list of priority projects. One state official told me how Cuomo's team did not rely solely on their charm. He bluntly said about the governor's top aide, that Schwartz "grabbed the county executives by the balls and squeezed until they cried uncle."

Streamlining Worked

In October 2011, the state and federal transportation agencies set a very ambitious schedule—to complete the environmental review in August 2012.[93] Remarkably, the agencies only missed their target date by a few weeks. When federal officials signed off on the ten-thousand-page EIS in September 2012, Governor Cuomo said, "This was the aspect of the project that had me holding my breath." He added, "This was really the big hurdle."[94]

Clearing that hurdle was a team effort that required direction, resources, time, and commitment. At various times under Pataki, Spitzer, and Paterson, state agency staff had not worked well together, lacked a sense of urgency, reviewed materials slowly, and were concerned about minimizing planning costs. They were bogged down trying to evaluate alternatives for a project that never was fully defined. After Cuomo and President Barack Obama expedited the review process, everything changed.

Once the project team was given clear direction from the governor's office, they were able to focus on completing the environmental review instead of arguing about the project's elements. While the project team prepared the EIS, the governor's office, agency heads, and communications staff dealt with the media, elected officials, and the public.

As soon as the White House selected the Tappan Zee Bridge project, the federal agencies developed a memorandum of agreement to define the agencies' roles and expectations. They committed to providing substantive comments with a short turn-around time.[95] The federal government also set up a multiagency "permitting dashboard" at www.permits.performance.gov to track the status of the review process. This website monitored the work of eight different federal agencies and included contact names, responsibilities, and status. It was like posting ongoing report cards. One federal highway official said, "We knew people were looking over our shoulders," and "everyone clicks their heels together when the president wants something."[96]

The governor's office put pressure on its agency heads to move the project along. In turn, state bureaucrats worked days, evenings, and weekends to please their bosses. Given the project's importance to the governor, spending money on the environmental review phase was not an issue. The state agencies did not challenge the consultants' expenses the way they typically did, whether it was for additional data collection, more analyses, or mailing documents via overnight mail.[97]

Given the project's high profile and urgency, the consultants went above and beyond the call of duty. One consultant told me that the governor's office had put the fear of God into the project team: "We felt like we had to get it done." AKRF, the consulting firm hired to complete the environmental review process, started working in August 2011. By the end of 2011, even though they had already spent millions of dollars and assigned eighty staff members to various tasks, AKRF still had not signed a contract with the state.[98]

The agencies could not meet the governor's timetable if they prepared the environmental documentation in their usual manner. There was not enough time to wait for NYS DOT and the Thruway Authority to review materials before they were reviewed by all the other relevant state and federal agencies, so they set up a simultaneous review process. The consultants shared draft versions of documents directly with state and federal regulatory agencies so that potential issues could be identified early on. Since so many agencies were involved at both the state and federal levels, concurrent review saved months if not years. At meetings described as long, grueling, and heated, agency staffers who were authorized to make decisions sat down with the consultants and revised documents together, line by line. Some of the project team's sessions had as many as thirty participants (in person and via videoconference), including representatives from the Thruway Authority, the Federal Highway Administration, bridge design consultants, AKRF, NYS DOT's environmental specialists, and attorneys.[99]

As head of the consulting team, AKRF's Robert Conway turned the tables on the normal client-consultant relationship. When necessary, he would "yell" at government agency representatives when they moved too slowly.[100] AKRF's project managers did not treat the Thruway Authority and the NYS DOT staff as their bosses; instead they considered the project to be in charge. One team member said, "AKRF didn't mess around because they didn't want DOT to make them look bad." He explained, "Conway can be annoying; he can be a pain, but he gets to the point." He added that Conway has "tremendous self-confidence. He can be rude, but he knows his stuff."[101]

Referring to the meetings in which agency staff and consultants reviewed documents line-by-line, Conway himself said, "When you're on page six of a thousand-page document and it's been three hours, someone has to play the bad guy."[102]

AKRF's team could not have completed its work in less than a year if the original consulting team had not started its work years earlier. When AKRF picked up the draft EIS materials in the summer of 2011, about half the work had already been completed. AKRF was fortunate to have the other consulting firms still on board helping conduct additional evaluations relating to noise, air quality, water, and fish.[103]

When I interviewed consultants and government officials, I expected to find that they may have cut corners or ignored regulations in order to expedite the process. But that did not appear to be the case. They certainly worked harder and faster than usual. They clearly prioritized their Tappan Zee Bridge related work over just about everything else.

One official at a state regulatory agency said, "Arms weren't twisted and we didn't have to compromise our integrity." Another official explained that environmental reviews are often slowed down when project sponsors fight requirements to evaluate impacts. He noted that on the Tappan Zee Bridge project, regulatory agency officials promptly reviewed documents and the project team responded quickly to their questions, comments and concerns.

The project team was able to avoid unnecessary analysis. In 2009, the transportation agencies did not know how to deal with a request from a state environmental official who told the team to consider the project's impact on regional growth and significant development outside the thirty-mile corridor.[104] In 2012, with every agency knowing that the governor and president had made replacing the Tappan Zee Bridge a top priority, questions that sought information beyond the regulatory agencies' requirements got nipped in the bud quickly.

But expediting had a cost. One federal highway official said, "It's been tough on resources" because of overtime expenses and the staff's neglect of other projects.[105] Likewise, the governor and his senior aides had less time to focus on many of the state's other critical issues and projects.

Getting Construction Underway

In order to meet the governor's deadlines, Deputy Secretary Rae had to make sure the state simultaneously completed the planning process and selected a bridge contractor. She talked to Federal Highway Administration officials nearly every day about the environmental review and nudged the state's vast bureaucracy along. Under previous governors, the project team's executive steering committee met sporadically and often engaged in counterproductive finger pointing. Now, they were meeting about once a week and making decisions promptly.[106]

In November 2011, NYS DOT and the Thruway Authority requested detailed information from firms interested in designing and building the new bridge. The agencies quickly reviewed materials from the firms so they could identify a handful of qualified contractors who would be allowed to bid on the project. The governor's office was taking a risk because the state legislature still had not passed a bill authorizing the agencies to enter into a design-build contract.

In December 2011, Governor Cuomo received a present from the state legislature. In an action that was referred to as a "rush job pushed out in the year-end chaos in Albany," the legislature overwhelmingly passed a bill that had a little something for everybody including tax cuts, storm recovery funds, a job retention program, and authorization for the Thruway Authority to enter into a design-build project.[107]

The *Engineering News-Record* wrote that New York's first major design-build project, replacing the Tappan Zee Bridge, was on a "super fast-track," with state transportation officials and industry participants scrambling to meet Cuomo's push to have work under way by the end of 2012. One executive of a company preparing a response to the state's solicitation said he had been given more time to prepare for $50 million jobs.[108]

In March 2012, NYS DOT and the Thruway Authority issued a request for proposals to four qualified bidders. When the Thruway Authority selected the firm to design and build the new Tappan Zee Bridge in December 2012, Governor Cuomo stated, "This project was pushed, managed, cajoled every step of the way. What we were trying to do was impossible—to make this much progress on building a bridge in a year. It can take a year to buy a new desk in this place."[109]

Still, it was not too late for people to raise false expectations about the alternatives to Cuomo's bridge. The indefatigable Alexander Saunders told the Thruway Authority's board of directors that a combined rail and highway tunnel could be built faster and cheaper than a bridge, and that it would have no environmental impacts.[110] In December 2013, a new player took potshots at the state's plan. The billionaire Donald Trump claimed, "I could fix the Tappan Zee Bridge for relative peanuts. I could get it done for so little and with so little disruption that it would make your head spin."[111]

After more than thirty years of planning, a new Tappan Zee Bridge, as depicted in the accompanying figures, was finally about to become a reality.

FIGURES 11.2, 11.3, AND 11.4 (*above and facing page*) These 2014 renderings depict the new Tappan Zee Bridge. (*Source:* New York State Thruway Authority)

Rendering by Tappan Zee Constructors LLC and HDR Engineering Inc.

Rendering by Tappan Zee Constructors LLC and HDR Engineering Inc.

12

Lost Opportunities and Wasted Resources

If the New York metropolitan area's leaders had cooperated on implementing realistic improvements over the course of the last three decades, they could have dramatically improved the region's transportation infrastructure and services. Instead, they abandoned viable options, avoided politically controversial policies, and missed out on important opportunities. Most notably, they failed to construct a new rail crossing *under* the Hudson River.

Between the early 1990s and 2010, New York and New Jersey transportation agencies pursued two separate rail crossings across the Hudson River. While Metro-North, the New York State Department of Transportation (NYS DOT), and the New York State Thruway Authority studied the feasibility and conducted an environmental review for a new I-287 rail line, New Jersey Transit and the Port Authority followed a parallel path for the Access to the Region's Core (ARC) project. Unlike the I-287 rail project, however, construction of ARC actually commenced.

ARC involved building two new rail tunnels between New Jersey and midtown Manhattan, a new passenger concourse adjacent to Manhattan's Penn Station, and the connection for one-seat train service from Rockland and Orange Counties to Penn Station. In 2009 the two states were on their way to building the nation's most important public transportation project, the first new Hudson River rail crossing in one hundred years.

There is absolutely no doubt that the ARC project offered widespread benefits for the entire New York metropolitan area. The bus, rail, and vehicle crossings between New Jersey and New York City are at capacity levels during the peak period.[1] New Jersey Transit and Amtrak share two single-track rail tunnels between New Jersey and Penn Station that provided more than enough capacity in 1910 for intercity travel but were not designed to carry tens of thousands of twenty-first-century commuters who all want to get to work within an hour or two of each other. With more people commuting from New Jersey to New York than between any other two states, New York's U.S. Senator Charles Schumer says, "Getting into Manhattan from New Jersey is a nightmare."[2]

At the ARC's groundbreaking ceremony in 2009, the head of the Federal Transit Administration, Peter Rogoff, said, "We don't have the choice in not doing this project."[3] The megaproject would allow New Jersey Transit to double the number of peak-hour trains serving Penn Station.[4] It would improve service reliability, alleviate Penn Station's severe passenger crowding, and promote economic development on Manhattan's West Side. Amtrak customers traveling between Boston and Washington would benefit from more frequent and reliable service into New York. The Federal Transit Administration had awarded New Jersey a total of $3 billion from its nationwide competitive funding program, by far the largest amount a single project had ever been granted.

ARC offered far greater benefits than an I-287 rail line. Forecasters expected ARC to carry 254,000 passengers every weekday, more than nine times as many passengers as the 28,000 expected to use the I-287 rail line.[5] ARC would have cost more—$8.7 billion compared to $6.7 billion.[6] It is highly unlikely, though, that the Federal Transit Administration would have awarded the I-287 rail line any money from its competitive grant program because it ranked so poorly on its evaluation criteria.

As noted in previous chapters, transportation professionals throughout the region preferred ARC to rail along I-287. When the MTA's Peter Derrick was coordinating and comparing potential megaprojects, he told a Metro-North official that it was "insane" for the railroad to be thinking about building a new rail crossing when New Jersey Transit's service could serve so many more people.[7] At least one NYS DOT commissioner thought that the transportation agencies should build ARC to Manhattan and bus rapid transit along I-287. The Regional Plan Association's Jeff Zupan told the I-287 project managers, "It makes no sense for two separate agencies studying transportation solutions that address some of the same markets to effectively ignore the work of the other."[8]

Even though Metro-North's I-287 project manager, Janet Mainiero, recognized how important the ARC project was to the region, she kept quiet about it. She told me, "I kept my nose to the grindstone and I said to myself, this [I-287] is my project and the powers that be know better than me."[9] However, her boss, Howard Permut, criticized the ARC project both in private and in public.[10] He wanted to build a new east-west rail line so Metro-North would no longer need to rely upon New Jersey Transit for service west of the Hudson River. A New Jersey Transit official, Tom Schulze, understood why Permut wanted his own rail line to the other side of the river. Schulze said, "We're not always so easy. We put our customer's needs first. We have a good relationship with Metro-North, but we prefer control. Metro-North needs to go through us to change service, but our allegiance is to our riders."[11]

The MTA saw ARC as a competitor. MTA board member Ernie Salerno said the MTA was trying to get in front of ARC for federal money. He said, "We had our issues and they had theirs."[12] In a 2006 e-mail, an MTA executive frustrated by the slow pace of the I-287 process wrote, "Time is ticking by while the project has been unable to agree on whether or not it should even share information with electeds [elected officials] whose support is critical. Meanwhile the New Jersey ARC project has gained significant ground with Congress and the FTA [Federal Transit Administration] to our great disadvantage."[13]

New York's failure to work with New Jersey was terribly shortsighted. More than half the five hundred thousand daily suburban commuters to New York City cross the Hudson River.[14] New Jersey Transit service is critical to New York's corporations, which need to attract workers from the widest geographical area possible, and the metropolitan area's fastest-growing counties are located west of the Hudson River. Improved Amtrak service to Penn Station benefits New Jersey, but it helps New York even more. New York also would have benefited in another important way that many people do not realize; New Jersey residents who work in New York pay income taxes on their salaries to New York, not New Jersey.

As U.S. senator and New Jersey governor, Jon Corzine championed the ARC project, but less than a year after he lost the 2009 gubernatorial race, the new governor, Chris Christie, had a different perspective. He realized that ARC offered important regional benefits. However, he was the governor of New Jersey, not the region's czar or the nation's president. He wanted to repair his state's roadways without raising taxes on income, sales, or gas. So, much to the dismay of the region's transportation planners, in October 2010 he canceled the ARC project and used the money set aside by the state and the Port Authority for New Jersey highway projects instead.

In his announcement canceling ARC, Christie said it would have cost "far more than New Jersey taxpayers can afford."[15] Three weeks later, he rhetorically asked New Jersey town hall attendees "How much is New York City contributing? Zero. How much is New York State contributing? Zero. It's called the Access to the Region's Core, except the only people in the region who were paying for it were us."[16]

It took me a long time to understand why New York's governors and senior state officials were not outspoken in support of the ARC project—until a former NYS DOT commissioner explained to me, "If we said it's a good project then we would have had to help pay for it."

Turf Battles

The competition between ARC and the I-287 rail crossing was just one aspect of a larger turf battle. Over the past three decades, the nation's largest commuter railroads have fought over service territories and train stations in the New York metropolitan area. These political quarrels were directly and indirectly related to the I-287 project. They occurred on several battlefields and came at a great cost to the region.

One former MTA official refers to its two commuter railroads, Metro-North and the Long Island Rail Road (LIRR), as fiefdoms that often find themselves at loggerheads.[17] In 2012, Martin Robins, the director of Rutgers University's Transportation Center, said, "The MTA almost never tells its agencies what to do."[18] Referring to projects that require close coordination between the two railroads, one transportation official said, "There is a guerilla war going on."[19]

The MTA is now building a new rail connection that will bring LIRR service to Grand Central Terminal, the exclusive home of Metro-North. When the LIRR started planning this connection, its planners hoped to take advantage of Metro-North's unused platforms and tracks. But Howard Permut wanted to reserve excess capacity so he could expand Metro-North's services to areas north and west of its existing territory, including Rockland and Orange Counties. He told one of his planners, "Your job is to keep the Long Island Rail Road out of Grand Central."[20] The planner said to me, "It was unbelievable the excuses we came up with" to make that happen. (Metro-North's excuses were not totally unjustified; the LIRR project would have impacted Metro-North service during construction and required shoring up nearby tunnels and buildings.)

Instead of using Metro-North's tracks, LIRR trains will come into a new station below Grand Central Terminal in what is the largest mined underground terminal ever built in the United States. In 1996, the MTA estimated the cost of bringing the LIRR to Grand Central would be $2 billion and would be completed in 2012.[21] By 2014, cost estimates had risen to almost $11 billion, with the completion date pushed back to 2023.[22] A large portion of the cost overruns and delays have been related to building that deep terminal under Grand Central.[23]

Something very similar occurred when New Jersey Transit was planning the ARC project. Transportation planners hoped to connect their new tunnel with Penn Station and continue on to Grand Central Terminal. But in 2003, the planners reported that they were dropping the Grand Central option because of the need to obtain underground easements and the potential impact to Lexington Avenue subway services.[24]

Interviews with senior planners who worked on the study reveal a different interpretation of why they dropped the connection to Grand Central. A former New Jersey Transit planner said, "Permut had Metro-North dream up all sorts of excuses" why New Jersey Transit could not come in to Grand Central. Don Eisele, who worked at both LIRR and New Jersey Transit, said that

Metro-North claimed Grand Central Terminal was over capacity, even though it is the only two-level commuter railroad station in the world.[25] Grand Central has four times as many platforms as Penn Station, yet it accommodates less than half the number of trains each weekday.[26]

A former MTA planner, George Haikalas, offered another perspective about why the Grand Central connection was eliminated. He said New Jersey Transit's president, George Warrington, did not want the hassles of integrating his train services with other railroads. It was a similar impetus for Permut's plan to build a new east-west crossing. Warrington told Haikalas, "If you're not in control, you're fucked."[27]

Congestion Pricing

Economists promote congestion pricing as a way to improve the overall efficiency of a transportation network. This strategy was both easier and harder to implement on I-287 than on most other roads. It was easier because the Thruway and the Tappan Zee Bridge already had tolls. It was harder because after the bridge was built, the Thruway Authority started offering (and never removed) discounts to regular customers as a way to promote bridge traffic—a perk that commuters came to see as an entitlement for living "on the other side" of the Hudson River.

New York's transportation agencies have long understood the benefits of congestion pricing. In the 1980s, the New York Metropolitan Transportation Council determined that raising tolls on the Tappan Zee Bridge during the peak period would be one of the most effective measures to reduce traffic, and every Thruway Authority executive director since the 1980s has realized that giving steep discounts to regular commuters runs counter to NYS DOT's desire to reduce peak period traffic volumes.

However, the world is not run by economists. Although the Thruway Authority was set up as a public corporation with an independent board, the governor's office has long made sensitive policy decisions, and there is no more sensitive and highly charged decision than raising tolls. The idea of raising peak-period tolls on the Tappan Zee Bridge was discussed in every governor's office from Mario Cuomo to Andrew Cuomo, but none of them were ready to implement congestion pricing (except for commercial vehicles) or eliminate the Tappan Zee Bridge commuter discount. Governor Pataki said, "It's difficult to do when it affects someone's livelihood."[28]

The authority chairmen have been loath to raise peak period tolls. Virgil Conway, the chairman of the MTA and Governor Pataki's I-287 Task Force, said congestion pricing is "politically very difficult. You're really punishing the working man; it's a middle-class tax." Lou Tomson, a Thruway Authority chairman, described it as "a problem politically; people need to use the road when they have to use it. It's easier to raise tolls overall."[29]

Other important players in the I-287 planning process were also against congestion pricing. The Automobile Club of New York did not want its members to pay higher tolls, and Ross Pepe's industry coalition wanted projects that created construction jobs.[30] James Yarmus, Rockland County's planning and transportation commissioner, said congestion pricing was a "nonstarter" since "transportation is a right, like the right to breathe."[31]

In 2015, regular commuters with E-Z Pass traveling during the peak period paid $3 round-trip to cross the Tappan Zee Bridge and $11.75 to cross the George Washington Bridge. When I mentioned to a Federal Highway Administration official that it would be hard for the politicians to double or triple the Tappan Zee toll in order to reduce congestion during peak periods, he said, "You would have to raise it much more than that."[32]

The HOV Lane

Congestion pricing was just one element of a transportation demand management program that could have been implemented along the I-287 corridor. New York also missed the opportunity to build a Cross Westchester Expressway HOV lane that would have encouraged more ride sharing and transit use, and accommodated many more travelers.

If the lane had been built and remained dedicated to HOV users, it had the potential to become an important regional transportation resource—the foundation of an HOV network across Westchester, the Tappan Zee Bridge, and Rockland County. Although state officials were not considering it at the time, the Cross Westchester HOV lane eventually could have become part of an even larger network of HOV lanes that stretched from Long Island through New York City into the Hudson Valley and on to New Jersey. The New York region could have had an HOV beltway around Manhattan.

State officials initially planned on restricting the I-287 HOV lane to vehicles with two or more passengers, but as the use of the lane grew they could have increased the minimum number of passengers and even turned it into a bus-only lane. However, no HOV or bus lane will be built along I-287, since the state rebuilt the highway and its overpasses in Westchester without sufficient space in the median to accommodate one. Although it is possible that the state could restrict an existing lane to high-occupancy vehicles, NYS DOT officials have recognized since the 1980s that it is not politically feasible.

The Crumbling Bridge

The new Tappan Zee Bridge will make automobile travel faster, more reliable, and safer. It did not have to be replaced, however.

The graceful George Washington and Bear Mountain Bridges are expected to span the Hudson River for centuries; the Tappan Zee Bridge won't see its sixty-fifth birthday. The three-mile-long bridge would not have deteriorated so rapidly if the Thruway Authority used a more robust design and protected its structural steel from roadway salt. Likewise, it would literally be standing on solid ground if the state had built it a few miles south where its foundations could have been driven into solid rock.

The reality is that replacing the Tappan Zee Bridge appears to have become a self-fulfilling prophecy. The Thruway Authority has long wanted to increase its capacity. When the bridge opened in 1955, it was not expected to carry more than 27 million vehicles per year.[33] It exceeded that level by the 1980s. In the 1970s and 1980s, the Thruway was unable to get support for building a second span parallel to the existing bridge. The authority's desire to increase the bridge's capacity was trumped by the concerns of others that a second span would be both too expensive and environmentally destructive.

In 1997, the Thruway Authority's executive director, John Platt, wanted to replace the bridge for numerous reasons, starting with capacity constraints, but also because of expensive maintenance requirements, the need to re-deck the roadway, and the opportunity to initiate a grand public-private partnership. He was able to build support for a new bridge with his dubious claims that the cost to repair the bridge would be about the same as the cost to replace it with one that was much wider.

The authority cut back on some of the capital projects recommended by Platt's predecessor and on some of the maintenance it would have performed if it expected the bridge to stay up for several more decades. Pouring a lot of money into a bridge that is getting replaced is not an efficient use of resources. Because the Thruway Authority decided to replace the bridge, it deteriorated faster than it otherwise would have. The bridge had to be replaced because Platt wanted to replace it.

Chris Waite, the former Thruway Authority chief engineer, felt that a new bridge would be more cost effective than repairing the existing span because it would need far fewer piers and they would be made of significantly stronger and more durable concrete. However, he also recognized that the Authority probably put less money into the bridge after it decided to replace it. "When maintenance folks know that a capital project is under design and will soon deal with the problems they have been battling for years," he said. "They often back down a bit and turn their attention and resources to other areas."[34]

Wasting Time and Money

Three decades of studies wasted an enormous amount of resources. In 2012, the Cuomo administration claimed $88 million had been spent on planning during the previous decade with little to show for it.[35] That was an overstatement, because the project team did conduct important analyses that were incorporated into the new bridge's planning and design process. But it was also an understatement because Cuomo's team did not include the costs associated with planning the HOV lane, Governor Pataki's task force, or all the time spent by state engineers, planners, accountants, and lawyers over three decades. If the Cuomo administration counted all those costs, they could have cited a figure that far exceeded $100 million.

The state incurred even more unnecessary costs. One of the reasons the Thruway Authority wanted to build a new bridge in the late 1990s was to avoid replacing the bridge's deck. However, the environmental review process took so long that the authority had to spend $300 million dollars to do exactly that anyway—after five-foot-wide holes started opening up along the length of the bridge.[36]

The metropolitan area also forfeited billions of dollars in federal transportation funds. New York was unable to use the $200 million earmarked in the 1991 federal transportation law for the I-287 HOV lane. In addition, Governor Christie returned the $3 billion grant for the ARC project—a prodigious sum that the state's congressional representatives, governor, and transportation officials had labored for years to obtain.

New York also lost one final opportunity that reflected its "kick the can down the road" mentality. Drivers will face dramatically higher tolls after the new bridge opens, and hundreds of millions of dollars may very well be diverted from other priorities, because the state failed to save any money to pay for a new bridge. Thruway Authority senior officials would have liked to put away funds for a new bridge, but they realized that such funds would have been used for other purposes. When the Thruway Authority was seen as flush with cash in the early 1990s, it was forced to take over responsibility for other highways and the state's canal system. A prudent homebuyer will save up money before building a new home, but politicians are rarely rewarded for raising tolls and taxes before they actually need it.

Conclusion

With minimal opposition and few regulations to hinder its progress, New York was able to plan and construct the 559-mile-long New York State Thruway in the 1950s expeditiously. The federal government only needed to prepare a twelve-page document and conduct one public hearing to finalize its review for the Thruway's Tappan Zee Bridge. Although many property owners living along the right-of-way opposed the project, they were characterized in the media as protecting their own interests at the expense of much-needed progress. Highways, at that time, were seen as solving problems, not causing them.

Starting in 1980, New York had a very different experience when it tried to address congestion along I-287. There is no single reason the state needed more than thirty years to finalize a transportation plan for a thirty-mile-long corridor. The saga was too complicated to blame a single person, institution, or factor for the state's failures. Year after year, decade after decade, seven different factors worked together in an insidious way to thwart New York's efforts. Conveniently, these factors form an acronym—FAILURE.

- Funding that was insufficient
- Adverse goals
- Interagency conflict
- Lack of leadership
- Uncertainty about the alternatives
- Regulations that were onerous
- Expectations that were unrealistic

Each of these factors is described in this chapter. Although the FAILURE acronym is a useful mnemonic, it is easier to understand the causes of delays and how they relate to each other by considering them in a different order, starting with the factor often considered the main culprit in delaying transportation projects—regulations that were onerous. The other factors relate to the state's inability to develop a consensus on a feasible alternative. The most important factor of all, lack of leadership, is described last.

Regulations

The National Environmental Policy Act, which requires full disclosure of a project's environmental impacts, delayed the planning process both directly and indirectly. In the 1990s, the New York State Department of Transportation (NYS DOT) needed several years to prepare an environmental impact statement for its proposed HOV lane. Leaders of the Tri-State Transportation Campaign took advantage of this lengthy review period to mobilize opposition, and when NYS DOT published its environmental findings, Tri-State used evidence from these documents to rally even more opponents. The National Environmental Policy Act slowed down planning once again after Governor George Pataki's task force recommended in 2000 that the state replace the Tappan Zee Bridge and build a new rail line. Federal officials required the project team to evaluate every feasible alternative, even though it had already done so in the 1980s and again in the 1990s.

The transportation agencies had difficulty finalizing their environmental review because they had to satisfy numerous regulatory agencies and comply with myriad federal laws, regulations, and executive orders, including those relating to air quality, wetlands, floodplains, environmental justice, endangered species, historic sites, and fish habitat. On top of that, New York had its own applicable laws, regulations, and agencies.

Although the project team found the regulations frustrating, costly, and time consuming, they did force the state to be more honest. A Metro-North planner complained to me that in the early 2000s the Federal Transit Administration took six months to review the computer model that would be used to forecast train ridership.[1] It is hard to argue that this review was unnecessary, however, given the railroad's own track record of manipulating numbers when the federal government was not looking over its shoulders, in the early 1990s.

Regulations were a major reason why the state took more than thirty years to finalize its plans, but a more fundamental problem was the state agencies' inability to develop a consensus on an affordable project with one another, let alone with local officials, business leaders, community groups, and environmental advocates.

Funding

In the 1980s and 1990s, the state could afford to build an HOV lane. It was a relatively prudent alternative based on a realistic assessment of alternatives and available resources. When Congress earmarked $200 million for the project, NYS DOT thought it had sufficient funds to complete it.

Replacing the Tappan Zee Bridge with a bridge twice as wide was another matter, however. It was affordable as long as the state was willing to dramatically increase the existing bridge tolls and tap into other potential funding sources. Building a low-cost bus rapid transit system also was feasible, although it would have required an ongoing source of funds to subsidize new bus services.

Governor Pataki's vision of building a thirty-mile-long commuter rail line that would tie together New York City, White Plains, Stewart Airport, and the existing north-south rail lines was just a pipe dream, though. The state did try to reduce the megaproject's cost by eliminating the connection to Stewart Airport and the cross-Westchester component. Even a scaled-back rail project that would connect only with the Hudson Line was unaffordable, however.

Adverse Goals

State officials had trouble developing a consensus because they were unable to balance a multitude of adverse or conflicting goals—such as supporting economic growth, accommodating more vehicles, reducing traffic, improving mobility, avoiding construction impacts, keeping tolls low, preventing further sprawl, and protecting the environment.

The best alternative for the corridor depended upon the project's goal. If the primary goal was to reduce travel time for single-occupant vehicles, the best alternative might have been to add more highway lanes. If it was more important to protect the region's dwindling open space and accommodate twice as many commuters, creating transit-oriented developments would have been a far better choice.

Numerous individual and institutional goals conflicted. NYS DOT Commissioner White wanted to be a national leader in building HOV lanes, while Thruway Authority engineers resisted implementing the idea because they thought it would embarrass them. The Thruway Authority's John Platt and Metro-North's Howard Permut promoted ambitious projects that would catapult their careers and cement their legacies, while other senior officials played it safer and just tried to hold onto their jobs.

The governors' goals often conflicted with those of the bureaucrats and the public. Thomas Dewey's focus on saving time and construction costs resulted in a bridge with costly maintenance expenses. Andrew Cuomo's short-term focus made replacing the bridge a priority but it came at a cost to those who prioritized public participation, transparency, and transit.

Decision makers can overcome adverse goals, but not when they try to please everybody. State officials were unable to simultaneously satisfy two county executives, let alone a host of communities, institutions, and special interests.

Too much public involvement and sensitivity to every interest group was time-consuming, expensive, and counterproductive.

Uncertainty About the Alternatives

Generating a consensus was difficult not only because of a lack of funding and adverse goals, but also because state officials and the public lacked reliable information. Instead of crystal balls on their desks, the project team had to work under dark clouds of uncertainty.

Since planners cannot accurately predict demographics and future travel patterns, no one really knows how many people will use an HOV lane, a new rail line, or a bus route. Predicting future public policies relating to various transportation alternatives is also fraught with uncertainty. An HOV lane that opens in April may become part of a future regional bus lane network, or it could be converted into a general-purpose lane by Election Day. Moreover, no one really knows whether communities will decide to implement regulations that promote sprawl or transit-oriented development.

Cost estimates were unreliable, and they changed as transportation agencies learned more details about construction techniques, soil conditions, property acquisition needs, underground streams, and existing utility lines. State officials also had to take into account potential events far beyond their control. Would a terrorist plant a bomb on a vulnerable portion of the Tappan Zee Bridge? Would a passing ship accidentally hit the bridge's piers? Would an earthquake weaken the bridge's foundations?

Forecasting future conditions and planning improvements for a rapidly changing region that is home to over 22 million people is difficult enough. Certain state officials made the task much more difficult by manipulating data and hiding information that created a cycle of false expectations and even more uncertainty.

Expectations

The public and local officials seemed to think the state should provide congestion-free roads and offer convenient and inexpensive new rail services—all without raising tolls and taxes. Very few people understood the real costs, benefits, and feasibility of the various alternatives.

How could anyone be expected to develop an informed opinion when false expectations about the various alternatives were raised by elected officials, construction firms, civic advocates, business interests, and the press? These players filled the gaps caused by the uncertainty, and some of them exploited it for their own advantage.

Key players had false expectations based on misinformation. Governor Pataki was in the dark about the extraordinary challenges of building an east-west commuter rail line. Attorney General Eliot Spitzer was not privy to accurate information when he made campaign promises. When state officials and MTA executives looked for advice about a potential east-west rail line, they relied upon a Metro-North official who regularly twisted facts to promote his own agenda. Likewise, the boards of directors for the Thruway Authority and the MTA did not have accurate information on which to make their decisions.

Metro-North's study in the early 1990s deliberately underestimated the costs, overestimated the benefits, and failed to address the rail project's enormous complexities. Governor Pataki's task force did the same thing in 2000. When state officials recommended commuter rail in 2008, they massaged the data to place the commuter rail component in a more favorable light. By manipulating information, the state raised false expectations about the project and made it more difficult to generate consensus on a realistic plan.

The way officials distorted their analyses to show certain results is not an unusual phenomenon. The Danish researcher Bent Flyvbjerg finds that planners and promoters deliberately and strategically misrepresent forecasts in order to increase the likelihood that their projects, and not those of their competitors, gain approval and funding. He claims that strong incentives and weak disincentives have taught planners and promoters that "lying pays off."[2] This behavior reverberates throughout an organization. Martin Wachs, a highly regarded transportation researcher, finds that planners, engineers, and economists often revise their forecasts to satisfy their superiors.[3]

It is important to understand that not everyone who raised false expectations during the I-287 planning process did so deliberately. Researchers who study decision making find that people often use ideology as a shortcut and tend to misperceive actual levels of risk. These two forces help explain why Pataki, an optimistic governor with a political philosophy favoring public-private partnerships, thought the private sector could finance a commuter railroad.

False expectations were also caused by ignorance. Although many people were ready to offer their opinions, few were interested in taking the time to understand the complex aspects of the transportation system such as the operational constraints of railroads, the methodology the federal government uses to fund transit projects, or the inextricable relationship between land use and transportation.

Public hearings and meetings designed to promote public involvement exacerbated these false expectations. Reporters liked covering the project's public meetings because they offered good sound bites and visuals right before deadlines. Since the project team did not quickly try to shut down impractical concepts like the highway tunnel, the media gave relatively equal weight to everyone's ideas.

Transportation agencies also had false expectations about each other's intentions and resources. In the 1980s and 1990s, senior NYS DOT officials mistakenly thought the Thruway Authority would support the Cross Westchester

Expressway HOV lane, raise tolls during the peak period, offer larger discounts for carpoolers, and convert the seventh bridge lane to an HOV lane. In the 2000s, MTA officials mistakenly thought the Thruway Authority would help pay for a new railroad while Thruway Authority officials had false hopes that the MTA would help pay for a new bridge.

The flourishing of false expectations and ignorance both inside and outside government put participants in the planning process under pressure. Virgil Conway, the chairman of Pataki's task force in 2000, proposed an impractical set of recommendations because he was trying to satisfy the false expectations of the governor, the advisory committee, and the public.

Interagency Conflict

The complex regulations, insufficient funds, adverse goals, uncertainty, and false expectations created an ideal breeding ground for interagency conflict. Planning was repeatedly delayed when government officials were unable to work together or went behind each other's backs.

Although over fifty federal, state, and local agencies had a stake in the project, the most dysfunctional element involved the three powerful state transportation agencies that battled each other about transportation demand management measures, a rail line, HOV lanes, and a new bridge. In the 1980s, the conflict between NYS DOT and Thruway Authority stymied the state's ability to address congestion along I-287. In the 1990s, the Thruway Authority's opposition to NYS DOT's HOV plan emboldened the opposition. In the 2000s, Thruway Authority and Metro-North officials were barely on speaking terms when Howard Permut's holy grail became the Thruway Authority's ball and chain. The I-287 megaproject had indeed become the poster child for interagency conflict.

Sometimes attempts to avoid interagency conflict backfired. The state studied the rail line for too many years because NYS DOT commissioners and Thruway Authority executive directors were not interested in battling the MTA and Metro-North. Likewise, to avoid institutional conflict, Governor Spitzer designated Metro-North as the study's primary expert on transit issues, but that just led the state to recommend an unaffordable rail alternative.

Lack of Leadership

Given the multiagency, multi-jurisdictional nature of transportation improvements, New York's transportation agencies had difficulty choosing between different options for the I-287 corridor. The long planning process cried out for a leader who was able and willing to make a decision.

In the 1980s, the NYS DOT officials coordinating with their Thruway Authority counterparts had trouble deciding whether the Cross Westchester

Expressway should be six lanes, eight lanes, or six lanes with an HOV lane. Two decades later, Metro-North and the Thruway Authority were unable to narrow down a list of fifteen potential alternatives to five.

NYS DOT commissioner Franklin White had the clout to make a decision about the HOV lane along the Cross Westchester Expressway. He generated enough support and resources to move the project along, but he did not stay in Albany long enough to see it through to completion. Nor was he powerful enough to force Thruway Authority officials to implement changes that were seemingly against their best interest. When White moved on to a new position in California, no effective champion replaced him for the HOV lane project.

Although governors could have made decisions that ended interagency conflicts and moved the megaproject along, they did not necessarily have an interest in doing so. Projects that take five or ten years to build are not so enticing to politicians who face reelection every four years. Sometimes studying and delaying provide greater short-term political benefits.

In 1997, Pataki was willing to make a decision to cancel the HOV lane. It was a political no-brainer to cancel an unpopular initiative and establish a task force to address the problem from a seemingly fresh perspective. However, once Pataki decided to pursue a rail project, he and his two successors found they had little to gain from lowering the public's high expectations; instead they were trapped by them. It was preferable to keep studying the problems rather than making decisions that would have engendered opposition and jeopardized their reelection opportunities. Taking a more active role would have required them to take the unpalatable steps of raising revenue and jettisoning unaffordable but popular project elements.

Even when the governors were willing to make decisions, they did not necessarily want to be associated with them. Pataki publicly kept his distance from the task force deliberations, and Cuomo avoided blame for the decision to eliminate transit from the new bridge. Likewise, all governors dissociate themselves from decisions about raising tolls and fares, although they go out of their way to take credit when they have opportunities to lower them.

It is often impossible to find out who makes decisions in state government, let alone their rationale. When I asked Tomson about decision making in the governor's office, he said, "My Dad used to joke that if an officer pulls us over, we'll all go into the back seat so he can't give any one of us a speeding ticket"[4] Two other senior officials offered a similar perspective. John Shafer, a former Thruway Authority executive director, said, "If the governor is smart, you don't know if a decision was the governor's."[5] Pataki's first deputy secretary, Maryanne Gridley, said the governor's staff did not document the rationale for their decisions because they did not want a written trace of them.[6]

Carrie Laney, who worked on the I-287/Tappan Zee project as both a Pataki aide and the Thruway Authority planning director, said that everyone, from local elected officials to the consultants, "all wanted leadership and someone to see the project through." She explained that elected officials were wary of taking a leadership role because they "did not want to be caught in a lose-lose situation." If elected officials promoted a new rail line, they might be seen as slowing down the entire project; if they suggested abandoning the rail line, they would be characterized as anti-transit. Elected officials also had little motivation in getting out in front of a governor who might change the plan or be unable to finalize it.[7] Laney and other politically connected players told me the same thing—that there was never a champion for the ambitious bridge-and-transit project because there never seemed to be a real project to champion.

When Leadership Can Overcome FAILURE

Governors can overcome the FAILURE factors when it is in their interest to do so.[8] They need to see tangible results during their term of office; otherwise, they will have little incentive to invest their time and political capital into championing a project.

On three occasions during the three-decade planning process, effective leaders identified short-term results and overcame daunting obstacles. They were not interested in conducting one study after another. They forged a consensus among key civic and business leaders despite insufficient funding, adverse goals, interagency conflict, high levels of uncertainty, onerous regulations, and unrealistic expectations.

When Dewey decided to construct a superhighway across the state, he built statewide support, fragmented the opposition, identified both short-term and long-term funding sources, overcame the Port Authority's jurisdiction, and personally made critical decisions about the route. Although Dewey was no longer governor when the Tappan Zee Bridge opened in 1955, he did attend Thruway ribbon-cutting ceremonies as various sections of the roadway were completed.

When Pataki's inner circle decided to reconstruct the Cross Westchester Expressway, they obtained sufficient funding and designed improvements in a way that expedited the environmental review. Pataki's task force brought the three transportation agencies together, then co-opted environmentalists and public transportation advocates by appointing them to an advisory group. Long before Pataki left office, drivers were able to see the benefits of the construction work as portions of the Cross Westchester Expressway were reconstructed.

Cuomo turned a $19.8 billion project into one that cost only $4 billion by eliminating climbing lanes and interchange improvements in Rockland County, as well as new transit infrastructure across the corridor. He also secured White House support and a $1.6 billion federal low-interest loan. Cuomo worked the phones, organized a public relations blitz, sent out his most senior aides to address local concerns, and worked with key industry and union leaders. Well before his reelection campaign kicked into high gear, he could point to visible construction progress and take credit for generating hundreds of new jobs.

Dewey and Cuomo were willing to devote significant resources to their megaprojects. The public works superintendent under Dewey referred to the Thruway as the governor's "pet project," and six decades later Cuomo's budget director said that the governor was "hell-bent" on building a new bridge.[9] Once these two governors decided to champion their megaprojects, they effectively sold their plans to the public and elected officials.

A troubling lesson from the I-287 planning process though is that the leaders who expedited projects did so by making decisions behind closed doors with little community input. Everyone supports transparency, public participation, and transit, but they all got in the way of New York's efforts to finalize its plans.

Dewey kept information about the bridge's location a secret for as long as possible and then gave local officials very little opportunity to influence the project. Pataki's senior state officials minimized public discussion about the Cross Westchester Expressway improvements to help expedite its approvals. Cuomo belittled the level of participation that his predecessors had encouraged, limited the number of people involved in decision making, and avoided discussing details about financing the multibillion-dollar megaproject.

Despite the Best of Intentions

In many ways, the outcomes of the three-decade planning process were exactly the opposite of what most participants set out to accomplish.

In the 1980s, NYS DOT officials proposed an HOV lane because they wanted to ease congestion, conserve energy, reduce pollution, and avoid the costs and environmental impacts of building a new bridge. In the 1990s, the Tri-State Transportation Campaign's environmentalists poured much of their resources into killing the HOV plan. They would find the replacement projects to be far less appealing, however. The state widened I-287 in Westchester for general traffic and is now replacing the Tappan Zee Bridge with a new bridge that is twice as wide.

Ironically, Tri-State promoted a bus rapid transit system along the corridor in 2012. Building a new bus lane is quite similar to building an HOV lane; after all, a bus is a high-occupancy vehicle. The HOV lane that Tri-State helped kill would have provided the foundation for the bus rapid transit system that the organization later supported.

A further irony of the state's efforts is that a rail project that required a new bridge became a bridge project without any rail. Pataki's decision to replace the Tappan Zee Bridge was spurred on by his desire to build an east-west rail line. If the state had not needed a new bridge to accommodate a rail line, it might very well have chosen to repair rather than replace the Tappan Zee Bridge.

The new bridge will not reduce overall highway traffic, nor will it encourage more transit use and ride sharing, because the state inadvertently set up the Thruway Authority with a mission that would conflict with some of the region's broader goals. In the 1970s, the authority wanted to twin the Tappan Zee Bridge to accommodate more vehicles and generate more toll revenue, but NYS DOT officials opposed the idea because they were more sensitive to environmental and community concerns. When NYS DOT tried to reduce highway congestion in the 1980s and 1990s, the Thruway Authority repeatedly thwarted NYS DOT's efforts.

The Thruway Authority's need to borrow billions of dollars to replace the Tappan Zee Bridge will lead to the greatest irony of this entire saga. The authority serves two powerful institutions with very different interests—the bond market and the governor's office. To satisfy the bond market, the authority needs to maximize its revenues. Governors, however, exert the opposite pressure; they want the authority to minimize its toll increases. The only way the authority can satisfy them both is to encourage more vehicles to cross the bridge. Therefore, some future Thruway Authority executives undoubtedly will try to open the emergency access lanes for toll-paying customers and widen portions of the six-lane Thruway that lead to the eight-lane bridge. As a result, a 1980 study designed to reduce the number of single-occupant vehicles and avoid the need to build a second span will result in a two-span bridge with even more vehicles.

At the End of the Day

Despite all the messiness associated with its planning, the new bridge at the Tappan Zee will be a welcome addition to the nation's transportation system. A crumbling, functionally obsolete structure will be replaced with two enduring ones that improve travel time, safety, and reliability. The bridge and the entire region's economy will be less vulnerable to ship collisions, earthquakes, malicious intent, marine borers, and hurricanes. The new bicycle path will attract crowds on weekends, and the wide spans will provide the Thruway Authority with sufficient space to maintain and operate a critical roadway properly.

However, the ribbon-cutting ceremony for this new bridge should also be an elegy for the old bridge and for the opportunities that were lost while government officials tried to address the region's transportation problems. The obstacles and delays that shaped the long planning process should not fade from memory; they can help us better understand and overcome the complex challenges associated with improving the nation's troubled transportation infrastructure.

Notes

Introduction

1 Milikowsky, *Building America's Future*, 15.
2 American Society of Civil Engineers, *2013 Report Card for America's Infrastructure*, http://www.infrastructurereportcard.org, 2013.
3 Milikowsky, *Building America's Future*, 15.
4 Federal Highway Administration, "Intercounty Connector," in *Interim Guidance on the Application of Travel and Land Use Forecasting in NEPA*, March 2010, http://environment.fhwa.dot.gov/projdev/travel_landUse/icc-case-study/icc-case-study.htm.
5 Virginia Department of Transportation, "Route 29 Charlottesville Bypass," http://www.virginiadot.org/projects/culpeper/rt._29_bypass.asp.
6 David Kocieniewski, "Thinking Beyond a New Tappan Zee," *New York Times*, July 24, 2009.
7 Smith, *Ad Hoc Governments*, 209.
8 Frug, "Beyond Regional Government," 1766–1767.
9 Jackson, *Crabgrass Frontier*, 140.
10 Kantor, "The Coherence of Disorder," 450; Holguin-Veras and Paaswell, "New York Regional Intermodal Freight"; Henry Peyrebrune, telephone interview with author, October 29, 2012.
11 Frug, "Beyond Regional Government," 1783–1784.
12 David Paterson, interview with author, Harlem, January 17, 2013; Ginger Gibson, "Christie Rips N.Y. Gov. Paterson's Suggestion to Split Cost of Tappan Zee Bridge Renovation," *Star-Ledger*, December 3, 2010.
13 Paaswell and Berechman, "Models and Realities," 96.
14 Sierra Club, "The Dark Side of the American Dream"; Frug, "Beyond Regional Government," 1783; Flyvbjerg, "Truth and Lies About Megaprojects," 15.

Chapter 1. The I-287 Corridor: From Conception to Congestion

1 Glaeser, *Urban Colossus*.
2 New York State Canal Corporation, "Canal History," *New York State Canals*, http://www.canals.ny.gov/history/history.html.

3 Ibid.
4 Jackson, *Crabgrass Frontier*, "Foreword."
5 Jeffrey Anzevino (Scenic Hudson Senior Regional Planner), published in NYS DOT, *Scoping Comments Report* (2009).
6 "486 Mile Express Highway Would Be World's Greatest, Governor Says at Liverpool," *Evening Leader*, July 11, 1946; "Dewey Starts State Thruway," *Evening Recorder*, July 11, 1946.
7 Ibid.
8 The New York State Department of Public Works, the predecessor of the New York State Department of Transportation, actually built the New York State Thruway on behalf of the Thruway Authority.
9 Winders, "Public Authorities in New York State."
10 New York State Temporary Commission, *Staff Report on Public Authorities.*
11 Panetta, *The Tappan Zee Bridge*, 54.
12 Leo Egan, "Authority Planned to Push Thruway; Bond Issue Studied," *New York Times*, February 14, 1950.
13 Joseph Ingraham, "Port Body Gives In on Thruway Span," *New York Times*, May 12, 1950; Ingraham, "Port Bridge Plan Blocked by Dewey," *New York Times*, May 7, 1950.
14 E-mail from Jameson W. Doig to author, January 22, 2014.
15 Ingraham, "Port Bridge Plan Blocked by Dewey."
16 Although the Tappan Zee area is often referred to as the widest portion of the river, it is the second widest, per Wolf, *Crossing the Hudson*, 186, and Federal Highway Administration, *Tappan Zee Hudson River Crossing Project Final EIS.* According to a state document, the widest part of the river is in Haverstraw Bay, http://www.dos.ny.gov/opd/programs/WFRevitalization/LWRP_Monitoring/Croton_Monitoring Report%20.pdf.
17 Ingraham, "Port Bridge Plan Blocked by Dewey."
18 Ibid.
19 For more information about the consideration of other bridge locations, see "Tappan Zee Span for Thruway Seen," *New York Times*, April 27, 1950 and Wolf, *Crossing the Hudson*, 186.
20 Laws of New York 1935, Chapter 869; Panetta, *The Tappan Zee Bridge*, 178 and 49.
21 Robert Knight, "History Notebook," *Journal News*, October 15, 1969 (article refers to December 14, 1936 report).
22 Panetta, *The Tappan Zee Bridge*, 54.
23 "Jolts Along the Thruway," *New York Herald Tribune*, December 7, 1950; Ingraham, "Thruway's Links on River Specified," *New York Times*, December 10, 1950.
24 "Thruway Agency Chided at Hearings," *New York Times*, November 29, 1950.
25 "Westchester Urged to Aid Thruway Fight," *New York Times*, December 18, 1950.
26 Panetta, *The Tappan Zee Bridge*, 61; Fein, *Paving the Way*, 215.
27 "Reporter Gets Thruway Post," *New York Times*, April 28, 1950; Fein, *Paving the Way*, 217.
28 The bridge is eight hundred feet north of the Port Authority line, per Wolf, *Crossing the Hudson*, 186.
29 Associated Press, "Tallamy Denies Bridge Will Destroy South Nyack," *Schenectady Gazette*, December 22, 1950.
30 Ingraham, "Thruway's Links on River Specified."
31 "Jolts Along the Thruway."
32 "Tallamy Denies Bridge Will Destroy South Nyack," *Schenectady Gazette*, December 22, 1950; Editorial, "Don't Forget the Bridge," *Journal News*, September 12, 1982.

33 "Jolts Along the Thruway"; "Nyack Areas Fears the Thruway Means Razing of 250 Buildings," *New York Times*, December 17, 1950.
34 "Tappan Zee Span for Thruway Seen," *New York Times*, April 27, 1950; "Opinion Is Divided on Thruway Plan," *New York Times*, December 23, 1950; "Bridge Views Sent to State By Trustees: Grand View Board Goes on Record As Opposing Span, Asks Delay to Find Out Facts," *Rockland County Journal* (newspaper clipping from Nyack Library's historical archives does not have a date).
35 "Tappan Zee Bridge Called Error," *New York Times*, January 2, 1951.
36 "Regional Board Asks Thruway Span Delay," *New York Times*, December 9, 1950.
37 "Tappan Zee Bridge Called Error."
38 Ibid.
39 Dan Greenbaum, e-mail to author, August 23, 2011.
40 "Moses Endorses Thruway Bridge," *New York Times*, January 5, 1951.
41 Ketchum, *Report on Application*, 4–8.
42 Wolf, *Crossing the Hudson*, 186–187; Ketchum, *Report on Application*, 8; Wolf, *Crossing the Hudson*, 189.
43 Leonard DePrima, telephone interview with author, March 15, 2011.
44 NYS DOT, Metro-North, and Thruway Authority, May 31, 2007.
45 Wolf, *Crossing the Hudson*, 183.
46 "Low Bid Received on Thruway Span," *New York Times*, April 17, 1953. The bridge design changed from tied-arch design to cantilever-truss.
47 Thruway Authority, NYS DOT, and Metro-North, *Alternatives Analysis for Rehabilitation and Replacement*, S-1.
48 Ibid., S-1–S-2.
49 Keith Giles, telephone interview with author, December 12, 2012.
50 DePrima interview.
51 Sy Schulman, telephone interview with author, September 18, 2011. Note that the highway did face fierce opposition from some affected property owners. For example, see Merrill Folsom, "Several Thruway Links Studied, Westchester Wants None of Them: Link to Thruways Irks Westchester," *New York Times*, January 7, 1954.
52 Ingraham, "Port Body Gives In on Thruway Span"; Madigan-Hyland Engineers, "Estimated Traffic, Revenue & Expenses of the NYS Thruway Authority," October 15, 1951.
53 "Text of Dewey's Annual Message to Legislature Urging Ethics Code for Public Officials," *New York Times*, January 7, 1954.
54 Panetta, *The Tappan Zee Bridge*, 76; 1950 and 1970 county populations, per U.S. Census Bureau.
55 Rockland County Planning Board, *Rockland County; River to Ridge*, table II-1.
56 Schulman interview.
57 Mary Kay Vrba, "The Top Tenants," *Westchester County Business Journal*, October 22, 2010, http://westfaironline.com/34330/the-top-tenants.
58 Richard Reeves, "Loss of Major Companies Conceded by City Official," *New York Times*, February 5, 1971.
59 Schulman interview.
60 Linda Greenhouse, "White Plains: 'Downtown' for All Westchester," *New York Times*, April 30, 1973.
61 NY Metropolitan Transportation Council, "Tappan Zee Corridor Study: Phase 1."
62 Garreau, *Edge City*, 39.
63 Sierra Club, "The Dark Side of the American Dream."
64 Penny Singer, "Employers Seek Ways to End I-287 Congestion," *New York Times*, January 28, 1990.

65 Beauregard, *When America Became Suburban.*
66 Danielson and Doig, *New York: The Politics*, 79.
67 Ibid.
68 Tri-State Regional Planning Commission, *1979–1980 Annual Work Program*, October 1979.
69 John Omicinski, "State Pondering Second Tappan Zee Bridge," *Nyack Journal News*, September 28, 1978.
70 Federal Highway Administration, NYS DOT, and NYS Thruway Authority, *Tappan Zee Hudson River Crossing Project Final EIS*, chapter 4.
71 Ferrandino & Associates, *Interstate 287 Monorail Feasibility Study*, 1.
72 "Statement by S. J. Schulman, President, The Westchester County Association, Inc., at a public information meeting, October 5, 1982, at Greenburgh Town Hall, re: studies of the Tappan Zee Bridge and its approaches," NYS DOT Archives.
73 Stanley Kramer, interview with author, New York, August 15, 2011.
74 NYCRoads, "Newburgh Beacon Bridge: Historic Overview," http://www.nycroads.com/crossings/newburgh-beacon/.
75 Rich Peters, note to author, December 11, 2012.
76 Gerald Cummins (chairman of NYS Thruway Authority), letter to F. Howard Zahn (Bureau of Environmental Analysis, New Jersey DOT), February 11, 1976. (This letter was related to the September 26, 1975 EIS for I-287 in New Jersey.)
77 Edward Hudson, "A Second Tappan Zee Bridge Is Mired in Uncertainty," *New York Times*, May 27, 1979; Cummins to Zahn.
78 Omicinski, "State Pondering Second Tappan Zee Bridge"; Hudson, "A Second Tappan Zee Bridge Is Mired in Uncertainty."
79 Winders, "Public Authorities in New York State," 103.
80 Ross Pepe (executive secretary, Construction Industry Joint Labor/Management Council of Lower Hudson Valley and Catskill Region), letter to William Hennessy (NYS DOT Commissioner), September 29, 1978.
81 Peters, "Minutes from Meeting with Thruway Authority and AAA on the TZB," June 6, 1980, NYS DOT Archives.
82 Hudson, "A Second Tappan Zee Bridge Is Mired in Uncertainty."
83 Omicinski, "State Pondering Second Tappan Zee Bridge."
84 Schulman, October 5, 1982 interview.
85 William C. Hennessy (NYS DOT commissioner), letter to State Senator John Flynn, March 7, 1980, NYS DOT Archives.
86 S. J. Horner, "Tri-State Planning: Is a Change Needed?" *New York Times*, February 1, 1981.
87 Danielson and Doig, *New York: The Politics*, 143.
88 Matthew Wald, "For TriState Agency an Uncertain Future," *New York Times*, August 26, 1979.
89 Michael Sterne, "For Planning Purposes, L.I. Wants to Be a Breakaway Province," *New York Times*, April 9, 1978.
90 Martin Gansberg, "New Jersey Journal," *New York Times*, August 1, 1982.
91 Downs, *Still Stuck in Traffic*, 8.

Chapter 2. Searching for Congestion Solutions (1980–1988)

1 Rich Peters, telephone interview with author, September 12, 2011.
2 NYS DOT, *Project P-130: Quarterly Progress Report of U.S. EPA. Status of Studies in the 1979 NYS SIP Revision,* Third Quarter 1980, NYS DOT Archives.

3 NYS DOT Region 8 Planning & Development Group, *Letter of Interest. Comprehensive Transportation System Management Assistance Program. Tappan Zee Bridge and Approaches,* February 1981, NYS DOT Archives.
4 Peters telephone interview, September 12, 2011.
5 Ibid.; NYS DOT Region 8 Planning & Development Group, *Letter of Interest*, 13.
6 Rich Peters, memo to files, "Meeting with Thruway Officials on TZB and Approaches Study," August 25, 1981, NYS DOT Archives.
7 Alan Bloom, memo to files, "Tappan Zee Study—Conclusion," March 25, 1983, NYS DOT Archives.
8 Westchester Journal, *New York Times*, August 22, 1982.
9 A. E. Dickson (Regional Director), letter to Eugene Levy (Assemblyman), September 13, 1982, NYS DOT Archives.
10 John Salak, "Reversible Lanes Urged for Bridge," *Westchester Reporter Dispatch*, October 6, 1982.
11 Sy Schulman, October 5, 1982, "Statement by S. J. Schulman, President, The Westchester County Association, Inc., at a public information meeting, October 5, 1982, at Greenburgh Town Hall, re: studies of the Tappan Zee Bridge and its approaches," NYS DOT Archives.
12 Richard Wolf, "Reversible Lanes on Tappan Zee Supported by Commuters, Officials," *Westchester Reporter Dispatch*, September 30, 1982.
13 "Tappan Zee Hearings," *Westchester Reporter Dispatch*, October 2, 1982.
14 Rich Peters, memo to files, "TZB and Approaches TSM Study," November 16, 1982, NYS DOT Archives.
15 Ibid.
16 J. L. Larocca (NYS DOT Commissioner Designate), letter to Edward K. Davies (Westchester County DPW Deputy Commissioner), April 14, 1983, NYS DOT Archives.
17 Henry Peyrebrune (NYS DOT Office of Public Transportation), memo to J. L. Larocca (Commissioner), May 19, 1983, NYS DOT Archives.
18 Howard Mann, interview with author, August 19, 2011; Peters telephone interview, September 12, 2011.
19 NY Metropolitan Transportation Council, "Tappan Zee Corridor Study: Phase III," II-1.
20 The non-carpool rate was fifteen dollars for twenty trips within thirty days. Prior to the toll increase in 1980, this commuter rate had not gone up since 1955. Carpoolers paid ten dollars for twenty trips within thirty-five days, per draft response of letter from NYS DOT Commissioner Hennessy to Senator Winikow prepared by Rich Peters on June 30, 1981, NYS DOT Archives.
21 NY Metropolitan Transportation Council, "Tappan Zee Corridor Study Phase III," II-2 (survey was conducted in 1983), NYS DOT Archives.
22 Rich Peters, interview with author, Poughkeepsie, February 18, 2011.
23 Carter Brown (Tri-State Regional Planning Commission), letter to Alan Bloom (DOT Regional Director), April 5, 1982, NYS DOT Archives.
24 NY Metropolitan Transportation Council, "Tappan Zee Corridor Study: Phase III," IV-81.
25 Note that interim reports were published by the New York Metropolitan Transportation Council but the final report was published by NYS DOT; NYS DOT, "Tappan Zee Corridor Study."
26 NYS DOT, "Preliminary Problem Definition and Project Proposal"; Federal Highway Administration and NYS DOT, *I-287/Cross Westchester Expressway: Final Design*

Report/Final EIS, II-36; A. E. Dickson, memo to T. G. Smith, re: draft design package, April 5, 1988, NYS DOT Archives.

27 Federal Highway Administration and NYS DOT, *I-287/Cross Westchester Expressway: Final Design Report/Final EIS*, II-37; NYS DOT, "I-287/Cross Westchester Expressway Development Study."

28 Peters telephone interview, September 12, 2011.

29 NYS DOT, "Preliminary Problem Definition and Project Proposal," 7.

30 Peters interview, September 12, 2011.

31 NYS DOT, "Preliminary Problem Definition and Project Proposal."

32 Ibid.

33 Rich Peters, memo to files, March 14, 1984.

34 Tim Gilchrist, interview with author, New York, October 18, 2012.

35 Andrew O'Rourke, letter to Commissioner Franklin E. White, January 7, 1988, NYS DOT Archives.

36 David Hechler, "New Office Prepares to Oversee Roadwork," *New York Times*, April 6, 1986.

37 Wayne Ugolik, e-mail to author, December 4, 2012.

38 W. H. Kikillus, memo to J. R. Lambert, September 8, 1988, NYS DOT Archives; Wayne Ugolik, interview with author, New York, December 13, 2012.

39 Charles Carlson (Deputy Commissioner for Departmental Operations), letter to James Martin (Executive Deputy Director of NYS Thruway Authority), September 23, 1985, NYS DOT Archives; "Initial Department Recommendations," April 7, 1986 (Signed by "Jeff," the last name was not legible), NYS DOT Archives.

40 Alan Bloom (NYS DOT R-8 Regional Planning and Program Director), letter to William Wheeler (New York Metropolitan Transportation Council Executive Director), August 23, 1985, NYS DOT Archives.

41 Alan Bloom, memo to E. W. Campbell (Planning Division in Albany), "Draft Recommendations of Tappan Zee Bridge Study," April 1986, NYS DOT Archives.

42 NYS DOT, summary pages and conclusion pages from "I-287/Cross Westchester Expressway Development Study PIN 8729.30," undated, NYS DOT Archives.

43 Ugolik interview.

44 NYS DOT, "Tappan Zee Corridor Study," 2.

45 The date the document was released was referred to in a letter from Franklin E. White to Congressman Benjamin Gilman on June 29, 1988, NYS DOT Archives.

46 NYS DOT, "I-287/Cross Westchester Expressway Development Study."

47 Lewis Hoppe, "Testimony to NYS DOT," submitted July 2, 1987, NYS DOT Archives.

48 NYS DOT, "NYS DOT, I-287/Cross Westchester Expressway Development Study"; Jerome W. Blood (Tarrytown's Deputy Mayor), letter to Albert Dickson (NYS DOT Regional Director), June 16, 1988, NYS DOT Archives.

49 NYS DOT, "I-287/Cross Westchester Expressway Development Study."

50 NYS DOT, "Meeting with Westchester County State Legislative Delegation," January 11, 1988, NYS DOT Archives.

51 NYS DOT, "I-287/Cross Westchester Expressway Development Study."

52 George E. Pataki, letter to Franklin E. White, July 13, 1988, NYS DOT Archives.

53 Henry Peyrebrune, telephone interview with author, October 29, 2012.

54 O'Rourke, letter to Franklin E. White, January 7, 1988, NYS DOT Archives.

55 Although the initial planning studies indicated that the eight-lane alternative required taking more properties than the HOV lane, NYS DOT's Final Design

Report/Environmental Impact Statement indicates the HOV lane and the eight-lane alternative would have required acquiring the same number of residential properties.

56 NYS DOT, "Meeting with Westchester County State Legislative Delegation," January 11, 1988, NYS DOT Archives.

57 O'Rourke, letter to Commissioner Franklin E. White, January 7, 1988, NYS DOT Archives.

58 Ibid.

59 Franklin E. White, letter to County Executive O'Rourke, December 22, 1987, NYS DOT Archives; per Rich Peters's December 11, 2012 handwritten note to author, the HOV lane would have been wide enough to convert into two general-purpose lanes or a light rail line for two reasons. First, it would have had wide shoulders so that disabled vehicles could pull out of the HOV lane. Second, it would have been physically separated from the general-purpose lanes to help enforce the HOV restrictions.

60 Rich Peters, memo to files, January 26, 1988, NYS DOT Archives.

61 Franklin E. White, letter to Congressman Benjamin Gilman, June 29, 1988, NYS DOT Archives.

Chapter 3. Finalizing Plans for the HOV Lane (1988–1995)

1 Franklin E. White, letter to Andrew O'Rourke, May 5, 1988, NYS DOT Archives.

2 Franklin E. White, draft of letter to task force with agenda, April 28, 1988 (target date), NYS DOT Archives.

3 Richard P. Duffy, "Remarks of Richard P. Duffy Presented to HOV/TSM (Transportation System Management) Task Force," June 22, 1989, NYS DOT Archives.

4 Ferrandino & Associates, *Interstate 287 Monorail Feasibility Study.*

5 Daniel Garvey, letter to Albert Dickson, November 2, 1988, NYS DOT Archives.

6 Monorail/Light Rail Subcommitee of the HOV Task Force, *An Assessment of Monorail/Light Rail Feasibility in the Suffern-Port Chester Corridor and Conditions / Features Required To Support Implementation*, undated, NYS DOT Archives.

7 Ibid.

8 Ibid.

9 Peter Eschweiler (Westchester County Commissioner of Planning), memo to Eric B. Langeloh (Commissioner of Transportation), April 26, 1989, NYS DOT Archives.

10 Robert Ancar, "Minutes of May 17, 1989 Management Briefing with F. E. White, HOV/TSM Task Force," May 23, 1989, NYS DOT Archives.

11 George Case, "HOVs: The Right Answer for 287 and Tappan Zee," *Westchester Environment* 89, no. 3 (October 1989), published by the Federated Conservationists of Westchester County, Inc.

12 HOV/TSM Task Force, "Draft Minutes," July 20, 1989, meeting, NYS DOT Archives.

13 Al Bauman, telephone interview with author, August 27, 2011.

14 HOV/TSM Task Force, "Draft Minutes," July 20, 1989, NYS DOT Archives.

15 Lou Rossi, telephone interview with author, November 12, 2012; Lou Rossi, memo to Henry Peyrebrune regarding meeting of April 12, 1988, with Thruway Authority, April 13, 1988, NYS DOT Archives.

16 HOV/TSM Task Force, "Draft Minutes," December 7, 1989, meeting, NYS DOT Archives.

17 W. Ugolik, memo to D. Erickson, January 25, 1989, NYS DOT Archives.

18 Wayne Ugolik, interview with author, New York, December 13, 2012.

19 HOV/TSM Task Force, "Draft Minutes," April 27, 1989, meeting, NYS DOT Archives.
20 Sy Schulman, "Westchester Opinion: The County Past, the County Future," *New York Times*. July 2, 1989.
21 HOV/TSM Task Force, "Draft Minutes," July 20, 1989, NYS DOT Archives.
22 Petrocelli et al., *I-287 Suffern-Port Chester Corridor HOV/TSM Action Plan.*
23 HOV/TSM Task Force, "HOV/TSM Newsletter," Spring 1989, NYS DOT Archives.
24 Petrocelli et al., *HOV/TSM Action Plan.*
25 Peters, telephone interview with author, September 12, 2011; Peters, note to Albert J Bauman, October 5, 1989, NYS DOT Archives; Al Bauman, note to Rich Peters, undated, NYS DOT Archives.
26 Editorial, "Think Monorail: More Roads Not the Answer," *Journal News*, April 27, 1989; editorial, "Our View: Moving People," *Journal News*, July 2, 1989.
27 "Marge vs. the Monorail," 1993 episode of *The Simpsons*, 20th Century Fox Home Entertainment, DVD.
28 Editorial, "Our View: Moving People."
29 HOV/TSM Task Force, *Operating Procedures*, June 23, 1988 NYS DOT Archives.
30 Alan Bloom, letter to Bernard Adler (City of White Plains Commissioner of Traffic), August 5, 1987, NYS DOT Archives.
31 William Kikillus (Regional Design Engineer) to Lou Rossi (Director, Planning Division), January 7, 1988, NYS DOT Archives.
32 William Kikillus, memo to J. R. Lambert, September 8, 1988 NYS DOT Archives.
33 NYS DOT, "I-287/Cross Westchester Expressway Development Study."
34 Kikillus to Rossi, January 7, 1988.
35 Winders, "Public Authorities in New York State," 187.
36 Ibid.
37 Federal Highway Administration and NYS DOT, *I-287 Final Design/Final EIS*, III-5.
38 Ugolik interview.
39 Peters interview, September 12, 2011; Senior Thruway Authority official, interview with author.
40 Henry Peyrebrune, interview with author, October 29, 2012.
41 William Demarest, "TZ Trolley Proposed: Commuters Present Plan to State DOT," *Journal News*, October 12, 1990.
42 Rich Peters, interview with author, Poughkeepsie, August 8, 2012.
43 Peters interview, September 12, 2011.
44 Federal Highway Administration and NYS DOT, *I-287 Final Design/Final EIS*, VI-53, III-7.
45 Peters, handwritten notes found with 1987 and 1988 documents, NYS DOT Archives.
46 Federal Highway Administration and NYS DOT, *I-287 Final Design / Final EIS*, viii.
47 Ibid., 1, A-2.

Chapter 4. Killing the HOV Lane (1994–1997)

1 George Case, "HOVs: The Right Answer for 287 and Tappan Zee," *Westchester Environment*, October 1989 (published by Federated Conservationists of Westchester County Inc.).
2 Ibid.
3 Jim Tripp, telephone interview with author, December 17, 2012.
4 Ibid.

5 Janine Bauer, telephone interview with author, August 19, 2011.
6 Tri-State Transportation Campaign, "Citizens Action Plan" (policy paper).
7 Bauer interview, August 19, 2011.
8 Ibid.
9 Ibid.
10 Tripp interview.
11 Tri-State Transportation Campaign, "First Public Edition of Mobilizing the Region," *Mobilizing the Region*, no. 7 (October 19, 1994).
12 Tri-State Transportation Campaign, "Reaching the Region," *Mobilizing the Region*, no. 59 (December 29, 1995).
13 Elsa Brenner, "State Invites Public's Ideas on I-287 Bottleneck," *New York Times*, November 19, 1995.
14 Michael Replogle comments in Federal Highway Administration and NYS DOT, *I-287/Cross Westchester Expressway: Final Design Report/Final EIS*, VI-14.
15 Bauer telephone interview with author, November 30, 2011.
16 Tri-State Transportation Campaign, "Questions for Westchester HOV," *Mobilizing the Region*, no. 45, August 3, 1995.
17 Bauer interview, November 30, 2011.
18 Bauer interview, August 19, 2011.
19 Tri-State Transportation Campaign, "Rough Road for Westchester HOV," *Mobilizing the Region*, no. 57, December 8, 1995; Tri-State Transportation Campaign, "NYS DOT Representative Addresses I-287 HOV Lane Proposal for Cross-Westchester and Tappan Zee," *Mobilizing the Region*, no. 29 (April 6, 1995).
20 Replogle, telephone interview with author, August 19, 2011.
21 Federal Highway Administration and NYS DOT, *I-287/Cross Westchester Expressway: Final Design Report/Final EIS.*
22 Replogle interview.
23 Regional Plan Association, "Jeff Zupan," http://www.rpa.org/users/jeff-zupan. Accessed July 24, 2012.
24 Jeff Zupan, interview with author, New York, December 22, 2011.
25 Statement by Scenic Hudson on December 6, 1995 in Federal Highway Administration and NYS DOT, *I-287/Cross Westchester Expressway: Final Design Report/Final EIS*, VI-66.
26 Federal Highway Administration and NYS DOT, *I-287/Cross Westchester Expressway: Final Design Report/Final EIS*, VI-20.
27 Zupan interview.
28 Ibid.
29 Maureen Morgan, interview with author, Ossining, February 8, 2011.
30 Ibid.
31 Morgan, telephone interview with author, January 2, 2012.
32 Ibid.
33 Morgan interview, February 8, 2011.
34 Federal Highway Administration and NYS DOT, *I-287/Cross Westchester Expressway: Final Design Report/Final EIS*, VI-306.
35 Morgan interview, February 8, 2011.
36 Morgan telephone interview, January 2, 2012.
37 Tri-State Transportation Campaign, "Major Westchester Paper Repudiates I-287 HOV Plan," *Mobilizing the Region*, no. 50 (September 8, 1995).
38 "Send HOV Proposal to Drawing Board," *Reporter Dispatch*, December 5, 1995.
39 William Casella, testimony at hearing and December 5, 1995 letter to Al Bauman in

Federal Highway Administration and NYS DOT, *I-287/Cross Westchester Expressway: Final Design Report/Final EIS, VI-111 and VI-443.*
40 Ibid., III-5.
41 Rich Peters, interview with author, Poughkeepsie, February 18, 2011.
42 Statement from Frank Ronnenberg in Federal Highway Administration and NYS DOT, *I-287/Cross Westchester Expressway: Final Design Report/Final EIS*, VI-467.
43 Statement from Audrey G. Hochberg in Federal Highway Administration and NYS DOT, *I-287/Cross Westchester Expressway: Final Design Report/Final EIS*, VI-96.
44 Sandra Galef, letter to Al Bauman, December 22, 1995 in Federal Highway Administration and NYS DOT, *I-287/Cross Westchester Expressway: Final Design Report/Final EIS*, VI-322.
45 Testimony of Naomi Matusow in Federal Highway Administration and NYS DOT, *I-287/Cross Westchester Expressway: Final Design Report/Final EIS*, VI-421.
46 Robert Weinberg, "HOV Lane Remains Best Possible Solution," *Reporter Dispatch*, January 17, 1996.
47 Federal Highway Administration and NYS DOT, *I-287/Cross Westchester Expressway: Final Design Report/Final EIS*, VI-451.
48 Brenner interview, November 19, 1995.
49 Peters interview, February 18, 2011.
50 Sarah Rios, interview with author, October 24, 2011.
51 Peters interview, February 18, 2011.
52 Al Bauman, telephone interview with author, August 27, 2011.
53 Ibid.
54 Morgan interview, February 8, 2011.
55 Morgan telephone interview, January 2, 2012.
56 Richard Newhouse (former Thruway Authority engineer), telephone interview with author, August 29, 2011; Keith Giles, interview with author, March 5, 2012; Stephen Morgan, interview with author, November 12, 2012; Duane Dodds, telephone interview with author, August 8, 2011.
57 Newhouse telephone interview, August 29, 2011.
58 Dodds telephone interview, August 8, 2011.
59 Leonard DePrima, letter to Al Bauman, December 26, 1995, in Federal Highway Administration and NYS DOT, *I-287/Cross Westchester Expressway: Final Design Report/Final EIS.*
60 Newhouse telephone interview, August 29, 2011.
61 Two former NYS DOT officials, two separate interviews with author.
62 Tri-State Transportation Campaign, "Around the Region: Federated Conservationist President Steps Down," *Mobilizing the Region*, no. 83, June 14, 1996.
63 Editorial, "O'Rourke Heading Wrong Way on HOV," *Gannett Suburban Newspapers*, January 8, 1996.
64 Andrew P. O'Rourke, letter to Ron Patafio (Gannet Suburban Newspapers Editorial Page Editor), January 10, 1996.
65 Ibid.
66 Peters, interview with author, Poughkeepsie, February 18, 2011.
67 Joseph Berger, "H.O.V. Lanes: A 30-Mile Test That Failed," *New York Times*, December 1, 1998; Joe Sharkey, "Speak Up If HOV Lanes Were Your Idea," *New York Times*, January 19, 1997.
68 Tri-State Transportation Campaign, "Road to Ruin Report Released—NY I-287 HOV, CT Rte. 6, and NJ Rte. 29 Cited," *Mobilizing the Region*, no. 131 (June 20, 1997); Tri-State Transportation Campaign, "Gannett Raps I-287 HOV

Project—Says Cross-Westchester Plan Is Road to Ruin," *Mobilizing the Region*, no.132 (June 27, 1997); Tri-State Transportation Campaign, "Westchester HOV Lane Faces Cascading Opposition," *Mobilizing the Region*, no. 138 (August 8, 1997).

69 Tri-State Transportation Campaign, "Traffic Alert," *Mobilizing the Region*, no. 143 (September 19, 1997); Tri-State Transportation Campaign, "Pataki Questions I-287 HOV Project," *Mobilizing the Region*, no. 147 (October 17, 1997).

70 Morgan, telephone interview, January 2, 2012.

71 Associated Press, "Resignation of Thruway's Chief Gives Pataki Ability to Revamp," *New York Times*, December 31, 1995.

72 Zupan interview, December 22, 2011.

73 Morgan interview, February 8, 2011; Bauer interview, August 19, 2011.

74 Maureen Morgan, e-mail to author, August 20, 2011.

75 George Pataki, interview with author, Rockefeller Center, New York City, December 14, 2011.

76 Bauer interview, August 19, 2011.

77 Pataki interview, December 14, 2011.

78 Andrew C. Revkin, "Pataki Cancels Project to Build Car Pool Lanes in Westchester," *New York Times*, October 21, 1997; Tri-State Transportation Campaign, "Governor Pataki Pulls Plug on Cross-Westchester HOV Plan," *Mobilizing the Region*, Special HOV Issue, October 20, 1997.

79 Revkin, "Pataki Cancels Project to Build Car Pool Lanes in Westchester."

80 Peter Samuel "New York State: Governor Ends Cross Westchester HOV," *Toll Road News*, December 1997.

81 Ross Pepe, telephone interview with author, November 30, 2012.

Chapter 5. Permut's Rail Line and Platt's Bridge

1 Patrick T. Reardon, "Burnham Quote: Well, It May Be," *Chicago Tribune*, January 1, 1992.

2 Doig, "If I See a Murderous Fellow," 300.

3 Isabel Wilkerson, "Cuomo Plan Would Let 3 Counties Quit M.T.A," *New York Times*, April 1, 1986; James Feron, "Rockland Votes to Withdraw from M.T.A," *New York Times*, March 6, 1988.

4 Jane Rosenberg, "Mid-Hudson, NYC Rail Study Set," *Times Herald Record* (Middletown, NY), May 5, 1989.

5 Marty Huss, telephone interview with author, March 7, 2011.

6 Ibid.

7 Matthew Wald, "Rail Link Would Cross the Hudson," *New York Times*, June 17, 1993.

8 Parsons Brinckerhoff, *Feasibility and Benefit-Cost Study.*

9 Thomas Parody, telephone interview with author, March 22, 2011.

10 Ibid.

11 MTA official, interview with author. (Note that "MTA" refers to the MTA headquarters staff and its transportation subsidiaries: Metro-North, Long Island Rail Road, and NYC Transit.)

12 Parsons Brinckerhoff, *Feasibility and Benefit-Cost Study.*

13 The more realistic assumption of seventy miles per hour would be used by other subsequent studies, such as page 3–22 of the report by Vollmer et al., "Final Report for Long Term Needs Assessment."

14 Jim Barry (MTA planner), telephone interview with author, July 11, 2011.

15 Parody interview.

16 Ibid.
17 Pickrell, "A Desire Named Streetcar."
18 Tessa Melvin, "Reaction Is Split on a New Hudson Rail Crossing," *New York Times*, January 6, 1991.
19 MTA official, telephone interview with author, February 3, 2011.
20 John T. Grant (Rockland County executive), letter to Louis Heimbach (Orange County Executive), August 4, 1988, in author's possession.
21 George Case, "HOVs: The Right Answer for 287 and Tappan Zee," *Westchester Environment* 89, no. 3 (October 1989); Melvin, "Reaction Is Split on a New Hudson Rail Crossing."
22 Peter Derrick, interview with author, Riverdale, New York, May 26, 2011.
23 MTA official, telephone interview with author, February 3, 2011.
24 Harold Faber, "Metro-North Studying a New Hudson Bridge," *New York Times*, October 13, 1990.
25 Federal Highway Administration and NYS DOT, *I-287/Cross Westchester Expressway: Final Design Report/Final EIS*, III-9.
26 R. Ancar (NYS DOT Downstate Metropolitan Planning), memo to file, "Minutes of May 17, 1989 Executive Management Briefing with F. E. White—HOV/TSM Task Force," May 23, 1989, NYS DOT Archives.
27 MTA official interview.
28 Maryanne Gridley (senior official in Pataki administration), telephone interview with author, August 5, 2011.
29 Tri-State Transportation Campaign, "Pataki Backs Port Authority JFK Rail Plan, Regional Rail Connections," *Mobilizing the Region*, no. 77, (May 3, 1996); Pataki, "Master Links: A Regional Transportation Vision for the 21st Century," undated, in author's possession (this report was prepared by the MTA with Governor Pataki's name on it, but it was never released to the public).
30 Gridley telephone interview.
31 MTA official interview.
32 John Shafer, telephone interview with author, September 8, 2011; Tony Gregory, telephone interview with author, November 12, 2012.
33 Gregory telephone interview; NYS Thruway Authority, Board of Directors meeting minutes, May 19, 1994.
34 Gregory telephone interview.
35 Winders, "Public Authorities in New York State."
36 Paul Muccino, "Earthquake Engineering Isn't Just for California Any More," *Roads and Bridges*, December 28, 2000, http://www.roadsbridges.com/earthquake-engineering-isnt-just-california-any-more.
37 Keith Giles, telephone interview with author, March 5, 2012.
38 NYS Thruway Authority, "The TZB Capital Program Coordination Committee's Report and Recommendation," June 1995.
39 John Shafer referred to Morgan as his longtime political advisor in interview with author, September 8, 2011.
40 Stephen Morgan, telephone interview with author, November 12, 2012.
41 Ibid.
42 Winders, "Public Authorities in New York State"; Marc Humbert for the Associated Press, "Pataki: There Will Be No Increase in New York Thruway Tolls," *Hour* (Norwalk, CT), December 5, 1996;
43 Tim Gilchrist, interview with author, New York, October 18, 2012.

44 Associated Press, "Ohio Official to Head Thruway Authority," *Albany Times Union*, October 3, 1996.
45 Senior governor's aide, interview with author,
46 Levy, *Build, Operate, Transfer.*
47 Lawrence DeCosmo, telephone interview with author, December 10, 2012.
48 NY State Thruway Authority, *General Revenue Bonds Series C*, February 1, 1995, i and 25.
49 Robert Dennison (former NYS DOT chief engineer), telephone interview with author, August 17, 2011.
50 Giles interview.
51 Richard Newhouse (former Thruway Authority engineer), telephone interview with author, August 29, 2011.
52 Former Paine Webber official, telephone interview with author, December 6, 2012.
53 Ibid.
54 NYS Thruway Authority, "Tappan Zee Charette," May 1 and May 2, 1997 (package of materials including memo, agenda, and briefing materials prepared for the two-day charette and obtained from the Thruway Authority's archives).
55 John Shafer, interview with author, September 8, 2011; Dennison interview.
56 Dan Greenbaum, telephone interview with author, August 15, 2011.
57 Dan Greenbaum, e-mail to author, August 23, 2011.
58 Lou Tomson, telephone interview with author, August 5, 2011.

Chapter 6. Pataki's Task Force: Raising Expectations Sky-High (1998–2000)

1 Quote from John Cahill, telephone interview with author, November 6, 2011.
2 James Dao, "Pataki, Keeping It Simple, Gains the High Ground," *New York Times*, January 5, 1998.
3 George Pataki, interview with author, Rockefeller Center, December 14, 2011.
4 Pataki and Paisner, *Pataki: An Autobiography*.
5 Pataki interview.
6 Virgil Conway, interview with author, New York, April 5, 2011.
7 Cahill telephone interview.
8 Tri-State Transportation Campaign, "Pataki Backs Port Authority JFK Rail Plan, Regional Rail Connections," *Mobilizing the Region*, no. 77, May 3, 1996; Pataki, "Master Links: A Regional Transportation Vision for the 21st Century," undated.
9 Cahill telephone interview.
10 Conway interview.
11 Ian Fisher, "Man in the News; For the M.T.A., a Pataki Loyalist: E. Virgil Conway," *New York Times*, April 22, 1995.
12 Pataki and Paisner, *Pataki: An Autobiography*.
13 Pataki interview.
14 Ibid.
15 Cahill, telephone interview.
16 Joe Sexton and Douglas Frantz, "A Powerful Fund-Raiser Who Also Oversees State Contracts," *New York Times*, September 3, 1995.
17 Andrew C. Revkin, "Pataki Cancels Project to Build Car Pool Lanes in Westchester," *New York Times*, October 21, 1997.
18 Tri-State Transportation Campaign, "Tackling the Tappan Zee Corridor, Post-HOV," *Mobilizing the Region*, no. 177 (June 19, 1998).

19 Ibid.
20 Robert Dennison, interview with author, August 17, 2011.
21 Ibid.
22 Tim Gilchrist, interview with author, New York, October 18, 2012.
23 Bill Varner, "New Tappan Zee Likely Fix for I-287," *Journal News* (White Plains, NY), October 8, 1999; Tri-State Transportation Campaign, "Governor Pataki Pulls Plug on Cross-Westchester HOV Plan," *Mobilizing the Region*, Special HOV Issue, October 20, 1997.
24 Jonathan Bandler and Jorge Fitz-Gibbon, "I-287 Overruns in Design, Inspections Push Project to $743 Million," *Journal News*, August 22, 2011.
25 Goetz, "Revisiting Transportation Planning," 264.
26 Vollmer et al., "Final Report for Long Term Needs Assessment."
27 Ibid.
28 Ibid.
29 Ibid., 2–12.
30 Tom Harknett, telephone interview with author, January 4, 2012.
31 Vollmer et al., "Final Report for Long Term Needs Assessment," 2-9.
32 Ibid.
33 Tri-State Transportation Campaign, "Westchester I-287 HOV Panned at Blue Ribbon Panel," *Mobilizing the Region*, no. 6 (October 12, 1994); Maryanne Gridley, telephone interview with author, August 5, 2011; Roberta Hershenson, "Going Beyond an Extra Lane on I-287," *New York Times*, May 7, 1995.
34 Lou Tomson, in August 5, 2011, telephone interview with author, talked about the great potential of guided buses and their ability to travel faster than regular buses. However, the buses would only travel at fifty-five miles an hour per Vollmer et al., "Final Report for Long Term Needs Assessment," 3–24.
35 Tri-State Transportation Campaign, "Thruway Completes Tappan Zee Toll Study," *Mobilizing the Region*, no. 238 (September 24, 1999); Resource Systems Group Inc., *Tappan Zee Congestion Relief Pricing Study—Final Report*, August 1999; Adler, Ristau, and Falzarano, "Traveler Reactions to Congestion Pricing."
36 The consultants' report used a range between $3.4 billion and $4.1 billion for the commuter rail alternative, but the task force's press release issuing its recommendation referred to the $4.1 billion estimate.
37 Vollmer et al., "Final Report for Long Term Needs Assessment," figure 4-2.
38 Pataki interview.
39 New Jersey Transit, *West Shore Corridor: Major Investment Study (MIS/DEIS) Draft Scoping Document*, June 2001.
40 Vollmer et al., "Final Report for Long Term Needs Assessment," 3-23.
41 Harknett telephone interview.
42 Federal Highway Administration and NYS DOT, *I-287 Final Design/Final EIS*, VI-242.
43 Dan Greenbaum, telephone interview with author, August 15, 2011; Harknett telephone interview; Vollmer Associates, "Strategies/Alternatives to be Dropped as a Result of Preliminary Screening," July 9, 1999, in author's possession.
44 Greenbaum telephone interview.
45 Vollmer Associates, "Strategies/Alternatives to be Dropped."
46 Ibid.
47 Ibid.
48 MTA, handwritten meeting notes titled "With Howard," August 2, 1999, in author's possession.

49 Harknett telephone interview.
50 Vollmer et al., "Final Report for Long Term Needs Assessment," 4–10.
51 Harold E. Vogt, letter to E. Virgil Conway, October 9, 1998, County Archives, Mount Vernon.
52 Bill Varner, "Morning Rush Hour Grows to 4," *Journal News*, June 14, 1999.
53 Conway interview, April 5, 2011.
54 I-287 Task Force, "I-287 Task Force Issues Recommendations and Study," press release, April 19, 2000.
55 David Chen, "State Task Force Recommends Replacing Tappan Zee Bridge," *New York Times*, April 20, 2000.
56 Harlem Valley Rail Trail Association, "History," http://hvrt.org/history_00.html.
57 Greg Clary, "Financial Questions Linger over Tappan Zee Plan," *Journal News*, May 1, 2000.
58 Tri-State Transportation Campaign, "MTA's T-Z Rail: A Pie-In-The-Sky?," *Mobilizing the Region*, no. 261, March 17, 2000.
59 Clary, "Financial Questions Linger over Tappan Zee Plan."
60 Conway interview.
61 Ray Rivera, "M.T.A. and Its Debt, and How They Got That Way," *New York Times*, July 26, 2008.
62 Rivera, "M.T.A. and Its Debt, and How They Got That Way,"; Permanent Citizens Advisory Committee to the MTA, *The Road Back: A Historic Review of MTA's Capital Program*, May 2012, http://www.pcac.org/wp-content/uploads/2014/09/The-Road-Back.pdf.
63 Richard Perez-Pena and Randy Kennedy, "Private Promoter for Transit Debt," *New York Times*, May 1, 2000.
64 Rivera, "M.T.A. and Its Debt, and How They Got That Way."
65 Clary, "Financial Questions Linger over Tappan Zee Plan."
66 Tri-State Transportation Campaign, "TZ II Rail Scrutinized at Citizen Forum," *Mobilizing the Region*, no. 275 (June 26, 2000).
67 Pataki interview.
68 Alex Philippidis, "Are the Counties' Roads on Track for the Year 2000?" *Westchester County Business Journal*, September 20, 1999.
69 Keith Giles, telephone interview with author, March 5, 2012.
70 Editorial, "Consider Higher Rush-Hour Car Tolls, Replacement Bridge for Tappan Zee," *Journal News* (White Plains, NY), September 27, 1999.
71 Tri-State Transportation Campaign, "Tappan Zee II on the Front Burner," *Mobilizing the Region*, no. 271 (May 29, 2000), emphasis added.
72 Bill Varner, "Pataki Comment of TZ Bridge Draws Fire," *Journal News*, July 17, 1999.
73 Chen, "State Task Force Recommends Replacing Tappan Zee Bridge."
74 David Chen, "Plan Urges New Bridge to Replace the Tappan Zee," *New York Times*, January 12, 2000.
75 Tri-State Transportation Campaign, "Vanderhoef Raises Big T-Z II Questions," *Mobilizing the Region*, no. 269 (May 15, 2000.)
76 Lou Tomson, telephone interview with author, August 5, 2011.
77 Marlin, *Squandered Opportunities*, 38; Pataki and Paisner, *Pataki: An Autobiography*.
78 Marlin, *Squandered Opportunities*, 36.
79 Conway interview.
80 Pataki interview.
81 Bruce Schaller, "Lessons From MetroCard Fare Initiatives," *New York Transportation Journal*, Fall/Winter 1998, www.schallerconsult.com/pub/metrocrd.htm.

82 Tomson telephone interview.
83 Richard Perez-Pena, "The Transit Showdown: Political Memo; Pataki's Transit Role? It's Behind the Scenes," *New York Times*, December 15, 1999.
84 Fisher, "Man in the News"; Conway interview.
85 Pataki interview.
86 As manager of planning at the MTA, I reported indirectly to Kupferman and worked with the task force's consultants.
87 In 1998, the MTA sold a South Bronx bus depot so that the land could be used by the *New York Post* for a new color printing plant. The newspaper, owned by Rupert Murdoch, provided critical support to New York's Republican candidates.

Chapter 7. The Thruway Authority versus Metro-North (2000–2006)

1 David Chen, "State Task Force Recommends Replacing Tappan Zee Bridge," *New York Times*, April 20, 2000.
2 Many of the people interviewed about Howard Permut's career and personality will remain anonymous.
3 Janet Mainiero, telephone interview with author, January 28, 2013.
4 Peter Melewski, telephone interview with author, February 26, 2013.
5 Peter Melewski, e-mail to Janet Mainiero, September 25, 2001, in author's possession.
6 Melewski, e-mail to Mainiero, October 4, 2001, in author's possession.
7 Ibid.
8 Melewski, e-mail to Mainiero, November 13, 2001, in author's possession.
9 Janet Mainiero, e-mail to Chris Waite, October 5, 2004, in author's possession.
10 Melewski telephone interview.
11 Ibid.
12 Mainiero, telephone interview .
13 Melewski telephone interview.
14 Leonard DePrima, telephone interview with author, March 15, 2011.
15 Mark Herbst, telephone interview with author, August 29, 2011.
16 Mainiero telephone interview.
17 Virgil Conway, telephone interview with author, February 3, 2011.
18 Melewski telephone interview.
19 I-287/Tappan Zee Bridge consultant, interview with author.
20 Chris Waite, telephone interview with author, August 13, 2011.
21 Mainiero telephone interview.
22 The Thruway Authority project managers who succeeded Peter Melewski were Charles Lattuca, Chris Waite, and Marc Herbst.
23 Will Ristau (NYS Thruway Authority), letter to Tom Schulze (New York Metropolitan Transportation Council), date stamped May [illegible date] 2001, in author's possession.
24 Chris Waite, e-mail to Michael Fleischer, August 25, 2003, in author's possession.
25 Robert Dennison, telephone interview with author, August 17, 2011.
26 Howard Permut, letter to Chris Waite, April 19, 2004, in author's possession.
27 Waite, letter to Permut, May 24, 2004, in author's possession.
28 Melewski telephone interview.
29 Mainiero, memo to file, re: telephone conference call with Thruway Authority on March 22, 2002, April 15, 2008 (note that the memo was dated six years after the phone call), in author's possession; Irwin Kessman, telephone interview with author, September 6, 2011.

30 In interviews with the author, Charles Lattuca and two MTA planners blamed federal officials. However, a Federal Highway Administration official thought that federal officials were not to blame for any delays; he thought the state was just trying to play off one federal agency against another in order to get the most favorable ruling.
31 Janine Bauer, letter to Andrew Spano, May 10, 2001, County Archives, Mount Vernon.
32 Ibid.
33 The terrorist attacks of September 11, 2001 also delayed the meeting because the New York Metropolitan Council's office was located on the eighty-second floor of the World Trade Center's Tower One. After the first plane hit Tower One, the council's offices were destroyed and three of its employees perished.
34 NYS DOT, "Notice of Intent for Tappan Zee Bridge /I-287 Environmental Review Published in Federal Register," press release, December 23, 2002.
35 Kreig Larson, "The Road to Streamlining," *Public Roads*, July/August 2003, http://www.fhwa.dot.gov/publications/publicroads/03jul/03.cfm; Federal Highway Administration, "Evaluating the Performance of Environmental Streamlining: Development of a NEPA Baseline for Measuring Continuous Performance," 2001, http://www.environment.fhwa.dot.gov/strmlng/baseline/section2.asp.
36 Greg Clary, "TZ Bridge Meeting Draws Calmer Input," *Journal News* (White Plains, NY), January 15, 2003; Greg Clary, "Spirits Run High at TZ Bridge Hearing," *Journal News*, January 16, 2003.
37 Greg Clary, "Plans Look at Easing I-287 Traffic," *Journal News* (White Plains, NY), March 27, 2003.
38 Many container ships do bring goods into New York City's Staten Island, but there is no direct rail connection between Staten Island and the rest of New York City.
39 Corey Kilgannon, "One Man's Vision: The Tappan Tunnel," *New York Times*, September 16, 2001.
40 Caren Halbfinger, "Rapid Buses Could Provide Ways to Curb Traffic Tieups," *Journal News*, July 11, 2003.
41 John Platt, letter to Andrew Spano, July 6, 2001, County Archives, Mount Vernon. (The Thruway Authority executive director told the Westchester County executive that federal and state laws prohibited him from precluding actions that would limit alternatives, since that would violate the impartiality of the alternative analysis process required under federal and state environmental laws.)
42 Greg Clary, telephone interview with author, February 1, 2013.
43 DePrima interview.
44 Caren Halbfinger, "Transit Officials Join to Seek Funds," *Journal News* (White Plains, NY), October 26, 2002.
45 Yilu Zhao, "Views of the Bridge," *New York Times*, July 6, 2003
46 As shown in figure G-2, Metro-North has three lines east of the Hudson River (Hudson, Harlem, and New Haven Lines) and two lines west of the river (Port Jervis and Pascack Valley Lines)
47 Greg Clary, "TZ Bridge Seen as Light Rail Span," *Journal News* (White Plains, NY), September 19, 2003.
48 Mark Kulewicz, e-mail to NYS Thruway Authority, March 4, 2003, in author's possession.
49 Halbfinger, "Rapid Buses Could Provide Ways to Curb Traffic Tieups."
50 Ibid.
51 Ed Buroughs, e-mail to Joyce Lannert and Lawrence Salley, February 6, 2001, in author's possession.

52 NYS Thruway Authority, Metro-North, "Long List of Level 1 Alternatives," June 8, 2003.
53 Caren Halbfinger, "Public Sees 15 Surviving Bridge Plans," *Journal News* (White Plains, NY), July 30, 2003.
54 NYS Thruway Authority, Metro-North, "Long List of Level 1 Alternatives," 12.
55 Halbfinger, "Public Sees 15 Surviving Bridge Plans."
56 Earthtech, "Level 1 Screening Charette Workshop," May 20–23, 2003, in author's possession.
57 The construction cost estimate to replace the long causeway and perform a full seismic retrofit was $1 billion, but the consultants added a 50 percent contingency and 35 percent for soft costs to come up with a $2 billion estimate. The same contingency and soft cost percentages were also used to estimate the costs of the other alternatives.
58 Greg Clary and Caren Halbfinger, "Cost of Replacing TZ Soars," *Journal News* (White Plains, NY), April 16, 2004.
59 Ibid.
60 Ed Buroughs, telephone interview with author, October 26, 2011.
61 Maureen Morgan, "Op-Ed; Metro-North Ignoring Needs of Orange, Rockland," *Times Herald Record* (Middletown, NY), July 20, 2004.
62 Clary and Halbfinger, "Cost of Replacing TZ Soars."
63 Greg Clary, "TZ Bridge Meetings to Start," *Journal News* (White Plains, NY), January 2, 2003.
64 Clary, "TZ Bridge Seen as Light Rail Span."
65 Clary and Halbfinger, "Cost of Replacing TZ Soars"; Greg Clary, "TZ Replacement Choices Now Due in September," *Journal News* (White Plains, NY), June 4, 2004.
66 Floyd Lapp, "Spanning the Tappan Zee," *Journal News* (White Plains, NY), May 19, 2005.
67 Dennison telephone interview.
68 Jeff Zupan, memo to Janet Mainiero and Chris Waite, April 28, 2004, in author's possession.
69 Chris Waite, memo to Janet Mainiero, July 7, 2004, in author's possession.
70 I-287/Tappan Zee Bridge consultant, interview with author.
71 Parsons Brinckerhoff, *Feasibility and Benefit-Cost Study*; Elaine Ellis, "Towns: Plan on Wrong Track," Gannett Suburban Newspapers, June 2, 1991; William Demarest, "Banning the Bridge at Both Ends," *Journal-News*, December 20, 1990.
72 Mainiero telephone interview; Waite, telephone interview with author, August 13, 2011.
73 Mainiero, e-mail to Waite, July 2, 2004; Mainiero, e-mail to Waite, July 7, 2004.
74 Mainiero, e-mail to Waite, November 18, 2004.
75 Mainiero, handwritten notes, March 4, 2005, in author's possession.
76 Waite e-mail to the following project team consultants on February 8, 2005: David Palmer (Arup), John Edy (Arup), Laurie Gutshaw (DMJM Harris), James Coyle (Earthtech), Tony Puglisi (Earthtech), Arnold Bloch (Howard Stein Hudson), in author's possession; Greg Clary, "Agencies Tie up TZ Corridor Plan," *Journal News*,(White Plains, NY), April 30, 2005.
77 Peter Cannito, memo to Michael Fleischer, April 7, 2005, in author's possession.
78 George Pataki, interview with author, Rockefeller Center, December 14, 2011. (Several interviewees from both the public and private sector also told me they had reached out to the governor's office.)
79 Sulaiman Beg, "2-County Task Force to Track TZ Bridge Plans," *Journal News*

(White Plains, NY), June 9, 2005; Andrew Spano, letter to William Fahey, July 6, 2005, in author's possession.

80 Pataki interview.
81 Robert Zerrillo, telephone interview with author, December 6, 2012.
82 Ibid.
83 Carrie Laney, telephone interview with author, November 28, 2011.
84 Marty Huss, telephone interview with author, March 7, 2011.
85 DePrima interview.
86 Virgil Conway, interview with author, New York, April 5, 2011.
87 Greg Clary, "TZ Bridge Proposals on the Way," *Journal News* (White Plains, NY), February 18, 2005.
88 Ibid.
89 Richard Newhouse, telephone interview with author, August 29, 2011.
90 Clary, "Agencies Tie up TZ Corridor Plan" (Dennison was referring to NYS DOT's predecessor, the NYS Department of Public Works, which built the Thruway on behalf of the Thruway Authority.)
91 NYS Thruway Authority, MTA Metro-North, and NYS DOT, "New York State Thruway Authority, MTA Metro-North Railroad, and New York State Department of Transportation Recommend Six Alternatives for Further Study in Tappan Zee Bridge/I-287 Environmental Review," September 29, 2005.
92 New York State Thruway Authority, Metropolitan Transportation Authority, and Metro-North Railroad, "Alternatives Analysis for Commuter Rail Hudson River Crossing Technical Memorandum," September 26, 2005.
93 I-287 project team consultant, interview with author.
94 NYS Thruway Authority, Metro-North, and NYS DOT, *Alternatives Analysis*, January 2006.
95 Khurram Saeed, "Lawmakers Briefed on Future of Tappan Zee Bridge," *Journal News* (White Plains, NY), September 19, 2006.
96 Caren Halbfinger, "Business Leaders Discuss Future of Tappan Zee," *Journal News* (White Plains, NY), November 3, 2006; Caren Halbfinger, "Tappan Zee Bridge Options Unveiled," *Journal News* (White Plains, NY), February 14, 2007.
97 Michael Anderson, interview with author, Tarrytown, March 8, 2011. Anderson referred to a fifteen-dollar toll.
98 Patrick McGeehan, "A Bridge That Has Nowhere Left to Go," *New York Times*, January 17, 2006.
99 Herbst telephone interview.
100 Mainiero, interview with author, Westchester, February 9, 2013.
101 Ernie Salerno, telephone interview with author, April 14, 2012.
102 Ibid.
103 Ibid.
104 Lou Tomson, telephone interview with author, August 5, 2011.
105 Ibid.
106 Pataki and Paisner, *Pataki: An Autobiography*.
107 James Dao, "Pataki, Keeping It Simple, Gains the High Ground," *New York Times*, January 5, 1998; Long Island Business News Staff, "Pataki Needs to Look to New York Skies," *Long Island Business News*, October 27, 2006.
108 Charles Lattuca, telephone interview with author, August 8, 2011.
109 MTA official, interview with author.
110 Former governor's aide, interview with author.

Chapter 8. Eliot Spitzer Doesn't Have Enough Steam (2007–2008)

1 Lou Tomson, telephone interview with author, August 5, 2011.
2 Ibid.
3 Eliot Spitzer, telephone interview with author, January 15, 2013.
4 Michael M. Grynbaum, "Spitzer Resigns, Citing Personal Failings," *New York Times*, March 12, 2008.
5 Steve Fishman, "The Steamroller in the Swamp: Is Eliot Spitzer Changing Albany? Or Is Albany Changing Him?," *New York*, July 14, 2007.
6 Eliot Spitzer, "Downstate Transportation Issues Speech: For Delivery at the Regional Plan Association's 16th Annual Regional Assembly," May 5, 2006, http://www.rpa.org/pdf/SpitzerRPATransportation.pdf.
7 Patrick Healy, "Spitzer and Clinton Win in N.Y. Primary," *New York Times*, September 13, 2006.
8 Spitzer telephone interview.
9 Michael Rothfeld, "Spitzer Vows to Fast-Track Rail Plans," *Newsday*, May 6, 2006, http://www.newsday.com/news/spitzer-vows-to-fast-track-rail-plans-1.549525.
10 Spitzer, "Downstate Transportation Issues Speech."
11 Ibid.
12 Ibid.
13 Ibid.
14 Rothfeld, "Spitzer Vows to Fast-Track Rail Plans."
15 David Paterson, interview with author, Harlem, January 17, 2013.
16 Elliot Sander, telephone interview with author, March 29, 2013.
17 Sewell Chan, "M.T.A. Director Calls for Ambitious Expansion," *New York Times*, March 3, 2008.
18 Tim Gilchrist, interview with author, New York, October 18, 2012.
19 Mary Ann Crotty, draft version of memo to Gilchrist, undated, in author's possession.
20 Robert Dennison, telephone interview with author, August 17, 2011.
21 Paterson interview.
22 Spitzer, "Downstate Transportation Issues Speech."
23 Paterson interview.
24 John Buono, letter to Andrew Spano and Scott Vanderhoef, February 22, 2007, in author's possession.
25 Andrew Spano and Scott Vanderhoef, letter to Governor Spitzer, February 14, 2007, in author's possession.
26 Ibid.
27 Spitzer telephone interview.
28 Marsha Gordon and Catherine Nowicki, letter to Michael Anderson, February 12, 2007, in author's possession; Gordon and Nowicki, letter to Spano and Vanderhoef, April 3, 2007, in author's possession.
29 Gordon and Nowicki to Spano and Vanderhoef.
30 Editorial, "More Overview Needed for TZ Overhaul Plans," *Journal News* (White Plains, NY), April 13, 2007.
31 NYS DOT senior official, interview with author.
32 NYS DOT, Thruway Authority, Metro-North, *Scoping Summary Report*, Appendix B and page D-1.
33 Neil Trenk, telephone interview with author, June 6, 2012; Naomi Klein, interview with author, Mount Vernon, New York, February 3, 2011; Mainiero, telephone interview with author, January 28, 2013.

34 Jeff Zupan, interview with author, New York, December 22, 2011.
35 Jeff Zupan, "Concerns about a Balance Discussion in the Alternatives Analysis Report," July 11, 2006, in author's possession; Jeff Zupan, memo to Janet Mainiero and Chris Waite, April 28, 2004, in author's possession; Regional Plan Association, Tri-State Transportation Campaign, Scenic Hudson, and Environmental Defense Funds, memo to Michael Anderson, May 4, 2006, in author's possession.
36 Tri-State Transportation Campaign, "Tappan Zee Symposium Showcases Bus Rapid Transit," *Mobilizing the Region*, no. 558 (June 15, 2007).
37 James Tripp, telephone interview with author, December 17, 2012.
38 Tom Schulze, interview with author, Newark, May 24, 2011; Sandy Hornick, telephone interview with author, December 19, 2011.
39 Rich Peters, interview with author, Poughkeepsie, February 18, 2011.
40 Paul Murphy, "A Warning from the Past on Hudson River Crossings," *Mobilizing the Region*, February 12, 2008; Michael A. Rockland, "The George: The Martha," *New Jersey Monthly*, October 6, 2008.
41 NYS Thruway Authority senior official, interview with author.
42 Elyse Knight, "Blind Approach to TZ Replacement," Opinion, *Journal News* (White Plains, NY), July 17, 2005.
43 Knight, telephone interview with author, March 9, 2011.
44 The NYS Thruway Authority referred to the fifty-year design life in its "2002 Annual Report" available at http://www.thruway.ny.gov/about/financial/ar/ar2002.pdf. In an August 8, 2011, telephone interview, Charles Lattuca (former Thruway Authority project manager) referred to the fifty-year design life. Also in a paper entitled "Tappan Zee Crossing of the Hudson River, New York," March Roche, the Thruway Authority's bridge consultant, referred to the fifty-year design life. (This paper was attached to e-mail from Roche to Mainiero on June 8, 2005.)
45 Knight interview; Khurram Saeed, "Thruway Debunks Tappan Zee Myth," *Journal News* (White Plains, NY), October 13, 2006; Michael Anderson, interview with author, Tarrytown, March 8, 2011.
46 Editorial, "That Awful, Vital Bridge," *New York Times*, January 22, 2006.
47 Khurram Saeed and Ken Valenti, "Westchester Environmental Groups Favor Trains as Part of New Tappan Zee Bridge," *Journal News*, February 12, 2008.
48 Ibid.
49 Bryan F. Yurcan, "TZ Conference Irks Task Force," *Westchester County Business Journal*, May 21, 2007; Tri-State Transportation Campaign, "Tappan Zee Bus Rapid Transit Symposium," *Mobilizing the Region*, no. 556 (May 22, 2007).
50 Marty Huss, telephone interview with author, March 7, 2011.
51 Carrie Laney, telephone interview with author, November 28, 2011.
52 Editorial, More Overview Needed for TZ Overhaul Plans, *Journal News* (White Plains, NY), April 13, 2007.
53 Editorial, "Pileup on the Tappan Zee," *New York Times*, March 4, 2007.
54 Judy Rife, "Citizen Groups Appointed to Advise Tappan Zee Study Team," *Times Herald-Record* (Middletown, NY), June 1, 2007.
55 NYS DOT, "State Charts New Course for Tappan Zee/I-287 Corridor Study: Streamlined Review and Increased Public Involvement Will Speed Completion," press release, January 17, 2008.
56 Based on Gilchrist memo to Spitzer, April 18, 2007, in author's possession, and author's interviews with project team members.
57 Governor Spitzer's Office, Draft of "Recommended Management Plan for Current

Tappan Zee I-287 Corridor Study and Completion of EIS/ROD Process," 2007, in author's possession.

58 Khurram Saeed, "State DOT Given Control Over TZ Bridge Project," *Journal News*, (White Plains, NY), May 18, 2007.

59 Gilchrist interview.

60 Governor's Office, Draft of "Recommended Management Plan for Current Tappan Zee I-287 Corridor Study and Completion of EIS/ROD Process."

61 According to a former senior governor's aide, Metro-North and the Port Authority were considered two of the most difficult agencies to control.

62 Cathy Woodruff, "Thruway Rejects Office Space for Bridge Project; Lease to Allow Agencies to Work Closer Together on Tappan Zee Replacement Deemed Too Expensive," *Albany Times Union*, May 17, 2007; New York State Thruway Authority, Board Minutes for September 19, 2007.

63 Senior Governor Spitzer official, interview with author.

64 Ibid.

65 Senior NYS DOT official, interview with author.

66 Khurram Saeed, "State DOT, Tappan Zee Bridge Task Force Discuss Vision, Purpose," *Journal News* (White Plains, NY), July 25, 2007.

67 Senior Governor Spitzer official, interview with author.

68 Sander interview.

69 Senior Governor Spitzer official, interview with author.

70 Mainiero, e-mail to Marty Huss, October 18, 2006, in author's possession.

71 Federal Highway Administration official, interview with author.

72 Gilchrist, memo to Eliot Spitzer, November 27, 2007.

73 Ibid.

74 Ibid.

75 Senior Governor Spitzer official, interview with author.

76 NYS DOT, "State Charts New Course for Tappan Zee/I-287 Corridor Study: Streamlined Review and Increased Public Involvement Will Speed Completion."

77 Saeed, "Federal Law Could Speed Tappan Zee Project," *Journal News* (White Plains, NY), February 4, 2008.

78 NYS DOT, NYS Thruway Authority, and Metro-North, *Scoping Comments Report (Tappan Zee Bridge/I-287 Corridor Project EIS.*)

79 Ellen Jaffe (Assemblywoman), Joseph Meyers (Rockland County Legislator), Dennis Kay (Mayor of Airmont), Jeffrey Oppenheim (Mayor of Montbello), and John Keegan (Mayor of Suffern), letter to Astrid Glynn, February 27, 2008 (in NYS DOT, NYS Thruway Authority, and Metro-North, *Scoping Comments Report*)

80 Khurram Saeed, "Bridge Review Process Flawed," *Journal News* (White Plains, NY), April 1, 2008.

81 Cesare Manfredi to Joan Dupont (NYS DOT Region 8 Director), April 13, 2008 (in NYS DOT, NYS Thruway Authority, and Metro-North, *Scoping Comments Report.*)

82 William Janeway (NYS Department of Environmental Conservation), letter to Michael Anderson, April 17, 2008 (in NYS DOT, NYS Thruway Authority, and Metro-North, *Scoping Comments Report.)*

83 Gilchrist interview.

84 Senior Spitzer official, interview with author.

85 Editorial, "Bridging the Divide Over the Tappan Zee," *Journal News* (White Plains, NY), August 5, 2007.

86 Gilchrist interview.

87 Spitzer telephone interview.

88 Ibid.

89 Alexander Saunders, "Tappan Zee Study Fails to Address Issues," letter to the editor, *Journal News* (White Plains, NY), May 29, 2008.

Chapter 9. David Paterson: The Overwhelmed Governor (2008–2010)

1 Robin Sidel, Dennis K. Berman, and Kate Kelly, "J.P. Morgan Buys Bear in Fire Sale, As Fed Widens Credit to Avert Crisis," *Wall Street Journal*, March 17, 2008; Ben White and Michael M. Grynbaum, "Life After Lehman Brothers," *New York Times*, September 15, 2008.

2 "Transcript of Governor David Paterson's State of the State address for 2009," *Post-Standard* (Syracuse, NY), January 7, 2009.

3 David Paterson, interview with author, Harlem, January 17, 2013.

4 Senior aide to the governor, interview with author.

5 Ibid.

6 Paterson interview.

7 Ibid.

8 Senior NYS DOT official, interview with author.

9 Tim Gilchrist, interview with author, New York, October 18, 2012.

10 Senior NYS DOT official, interview with author.

11 Governor's Office, "County Executive Briefing Talking Points: Opening Remarks," August 6, 2008, in author's possession.

12 Governor's Office, "County Executive Briefing Talking Points: Opening Remarks,"; NYS DOT, Metro-North, and Thruway Authority, PowerPoint, "TZB/I-287 Environmental Review County Executive Briefing," August 6, 2008, in author's possession.

13 New York State Department of Transportation, "Proposal for Tappan Zee Bridge & I-287 Corridor Unveiled: Team Recommends Bridge Replacement, Addition of Bus," press release, September 26, 2008.

14 Senior Paterson official, interview with author.

15 Paterson interview.

16 NYS DOT, press release, September 26, 2008 (emphasis added).

17 NYS DOT, NYS Thruway Authority, and Metro-North, *Transit Mode Selection Report*. A draft version of this report was issued in September 2008.

18 Ibid,, table 7–8. The table contains information about operating costs, number of transit users, and expected revenue for the various transit alternatives.

19 Khurram Saeed, "TZ Bridge Project to Bring in Planners to Work with Localities," *Journal News* (White Plains, NY), November 26, 2008.

20 Greg Clary, "Cost of Replacing TZ Soars," *Journal News*, April 16, 2004; Saeed, "New Assurances Offered at Forum on TZ Bridge," *Journal News* (White Plains, NY), December 14, 2005; Caren Halbinger, "Tappan Zee Bridge Options Unveiled," *Journal News* (White Plains, NY), February 14, 2007.

21 New York State Department of Transportation, "Tappan Zee Project Preliminary Financial Report Released: Lays Groundwork for Future Funding Plans," press release, November 20, 2008.

22 MTA official, interview with author.

23 Khurram Saeed, "Pilot Projects Could Lead to Private Funds for Tappan Zee Bridge," *Journal News* (White Plains, NY), June 3, 2009; Khurram Saeed, "Tappan Zee Bridge Team Ready to Unveil Routes for Trains, Buses," *Journal News* (White Plains, NY), June 13, 2010.

24 Tim Gilchrist, memo to Governor Spitzer, December 12, 2007.
25 Federal Transit Administration, *Annual Report on New Starts: Proposed Allocation of Funds for Fiscal Year 2007*, B-9.
26 Jeff Zupan, interview with author, New York, December 22, 2011.
27 Ibid.
28 Gilchrist, interview.
29 NYS Commission on State Asset Maximization, *Final Report.*
30 Gary Delaverson, telephone interview with author, March 7, 2012; Gilchrist interview.
31 Michael Anderson, interview with author, Tarrytown, New York, March 8, 2011.
32 Bob Baird, "Progress on new Tappan Zee Bridge Hasn't Limited Options, Issues," *Journal News* (White Plains, NY), June 27, 2010; Khurram Saeed and James O'Rourke, "Public Digs Into Tappan Zee Bridge Plan," *Journal News* (White Plains, NY), July 1, 2010; Khurram Saeed, "What You Might Have Missed About the Tappan Zee Bridge/I-287 Project at The Open House," *Journal News* (White Plains, NY), July 3, 2010; Bob Baird, "Winners, Losers Among Communities in Interstate 287-Tappan Zee Bridge Plan," *Journal News* (White Plains, NY), July 4, 2010.
33 J. Richard Capka explains how cost overruns in megaprojects can result from adding numerous small projects to a megaproject in his article "Megaprojects: They are a Different Breed," *Public Roads*, July/August 2004.
34 MTA official, interview with author.
35 NYS Thruway Authority, NYS DOT, and Metro-North, "Compilation of Base Data for DEIS Alternatives: Snapshot As of May 4, 2007," in author's possession.
36 Patrick Gerdin, interview with author, Pomona, New York, May 17, 2011.
37 Ed Buroughs, e-mail to Naomi Klein, March 24, 2009, in author's possession; Ed Buroughs, e-mail to Karen Pasquale, January 9, 2009, in author's possession.
38 TZB-I287 Rockland/Westchester Counties Working Group, "Draft Work Plan," January 23, 2009, in author's possession; Jerry Mulligan (Westchester County Planning Commissioner), letter to Michael Anderson, September 25, 2009, in author's possession.
39 NYS Thruway Authority, Minutes, Board Meeting No. 680, June 16, 2010; NYS DOT, NYS Thruway Authority, and Metro-North, "EIS Methodology Report," September 2010.
40 Council on Environmental Quality, Regulations for Implementing NEPA, Sec. 1502.7 Page Limits, http://www.gpo.gov/fdsys/pkg/CFR-2012-title40-vol34/pdf/CFR-2012-title40-vol34-sec1502–7.pdf.
41 NYS DOT, NYS Thruway Authority, and Metro-North, "EIS Methodology Report."
42 Ed Buroughs, e-mail to Kevin Plunkett, June 22, 2010, in author's possession.
43 Ibid.
44 Ibid.
45 Paterson interview.
46 Ibid.
47 Ibid.
48 Ibid.
49 Joseph Spector, "Paterson Turns to N.J. for New Tappan Zee Bridge," *Journal News* (White Plains, NY), December 2, 2010.
50 Ginger Gibson, "Christie Rips N.Y. Gov. Paterson's Suggestion to Split Cost of Tappan Zee Bridge Renovation," *Star-Ledger*, December 3, 2010.
51 Paterson interview.
52 *Saturday Night Live*, NBC television broadcast, September 25, 2010.

53 Richard Ravitch, *Report of the Lieutenant Governor.*
54 Charles J. Fuschillo, "Senator Fuschillo to Introduce Legislation Allowing New York State to Enter into Public-Private Partnerships to Expedite Road and Bridge Repairs Statewide," press release, April 15, 2011, http://www.nysenate.gov/press-release/senator-fuschillo-introduce-legislation-allowing-new-york-state-enter-public-private-p.
55 Gilchrist interview.
56 Richard Ravitch, interview with author, Rockefeller Center, February 28, 2012.
57 Ibid.
58 Ibid.

Chapter 10. Andrew Cuomo Takes Charge in 2011

1 Former Rockland County elected official, interview with author, 2011.
2 NYS DOT, Metro-North, and Thruway Authority, "TZB/I-287 Environmental Review: County Executive Briefing," August 6, 2008, in author's possession.
3 Ibid.
4 NYS DOT, Thruway Authority, Metro-North, "TZB/I-287 Corridor Project Newsletter—Autumn 2008," http://www.tzbsite.com/tzb-library/newslettershandouts/newsletter-autumn2008.html; NYS DOT, NYS Thruway Authority, and Metro-North, *Scoping Comments Report (Tappan Zee Bridge/I-287 Corridor Project Environmental Impact Statement)*; Thruway Authority, Metro-North, and NYS DOT, "Tappan Zee Bridge/I-287 Corridor Project: Fall 2010 Update," August 2010, http://www.tzbsite.com/public-involvement/trans-hwy-br-options/trans-hwy-br_pres201010/image1.htm.
5 Tri-State Transportation Campaign, "Governor Pataki Pulls Plug on Cross-Westchester HOV Plan," *Mobilizing the Region*, Special HOV Issue, October 20, 1997.
6 Khurram Saeed, "Tappan Zee Bridge Team Ready to Unveil Routes for Trains, Buses," *Journal News* (White Plains, NY)June 13, 2010.
7 County of Westchester, "Astorino Calls on Governor to Make TZB a Priority," press release, http://www3.westchestergov.com/news/all-press-releases/4054-astorino-calls-on-gov-cuomo-to-make-replacement-of-tz-bridge-a-priority; Robert Astorino, speech at Manhattan Institute's Forum on Replacing the Tappan Zee Bridge, June 22, 2011, http://www3.westchestergov.com/images/stories/pdfs/tzbspeech.pdf.
8 Ritchey, "Wicked Problems," 1.
9 Benjamin-Bothwell, "An Intervention Addressing a Wicked Problem," 30.
10 Conklin, *Dialogue Mapping*, 20.
11 Howard Milstein (Thruway Authority Chairman), letter to Thruway Authority board members with attached report from Navigant Financial Advisors, "New York State Thruway Authority Executive Summary Report," May 24, 2012, http://www.thruway.ny.gov/news/pressrel/letter-navigant.pdf.
12 Robert Weinberg, interview with author, Elmsford, NY, May 26, 2011.
13 Project team consultant, interview with author.
14 At that time, Metro-North's planners did not realize the governor's office was redefining the project to eliminate the rail component.
15 Project team consultant, interview with author.
16 MTA official, interview with author.
17 Jonathan Mahler, "The Making of Andrew Cuomo," *New York Times*, August 11, 2010; Celeste Katz, "Eliot Spitzer: Everyone Knows Andrew Cuomo's the Dirtiest, Nastiest Political Player Out There," *New York Daily News*, September 23, 2010.

18 David Paterson, interview with author, Harlem, January 17, 2013.

19 Muzzio, "Politics and the News Media," 201; Marquis Childs, "Governor Dewey Has Built a Highly Efficient Machine," *Pittsburgh Post-Gazette*, May 19, 1947; Jimmy Vielkind, "Comparing Cuomo to Dictator, GOP Pushes for Sunlight," *Times Union* (Albany, NY), February 5, 2013.

20 Paterson interview.

21 Jimmy Vielkind and James M. Odato, "Cuomo Control of Information Includes Screening, Redaction of Records at State Archives," *Times Union* (Albany, NY), July 23, 2012.

22 Jeremy Smerd and Daniel Massey, "Cuomo: The Boss. Nothing Gets Done in This State Without His Say. It's the Cuomo School of Management," *Crain's New York*, March 25, 2012.

23 Jacob Gershman, "Cuomo Digs, Schneiderman Parries," *Wall Street Journal*, August 15, 2011.

24 MTA senior official, interview with author.

25 Henry Peyrebrune, telephone interview with author, October 29, 2012, as well as an interview with another long-time senior NYS DOT official.

26 New York State Department of Labor, "State's Unemployment Rate Dropped to 8.3% in May, Lowest Since April '09," press release, June 17, 2010.

27 Andrew Cuomo, "Governor Andrew M. Cuomo State of the State Address," January 5, 2011, http://www.governor.ny.gov/s12/stateofthestate2011transcript. Accessed July 24, 2012.

28 Senior state official, interview with author.

29 On page 450 of his autobiography, *All Things Possible* (New York: HarperCollins, 2014), Andrew Cuomo recounts an early meeting in his administration at which he asked top construction and development people to identify an overdue transformative project that his administration could take on and complete.

30 Reporter, telephone interview with author, October 18, 2012.

31 Josh Robin, "Cuomo Continues Facilities Tour; Visits Tappan Zee Bridge," *NY1*, November 15, 2010, http://www.ny1.com/content/news_beats/128933/cuomo-continues-facilities-tour—visits-tappan-zee-bridge.

32 Based on review of e-mails in author's possession from Yomika Bennett, Suzanne Brackett, Michael Fleischer, and Linda Lacewell.

33 On May 20, the governor met with Lacewell, James Malatras, Bennett, Joan McDonald, Fleischer, and Joseph Martins. On August 4, the governor met with Larry Schwartz, Lacewell, Bob Megna, Bennett, Derek Utter, McDonald, Martens, and Mark Roche.

34 Tom Madison, Governor Andrew Cuomo's Cabinet Meeting, February 22, 2012, PowerPoint slides and video available at http://www.newnybridge.com/documents/meetings/index.html#cabinet (the Cuomo administration indicated that more than 430 public meetings were held during the previous ten-year period).

35 Tappan Zee Bridge/I-287 Environmental Review (Project Team), "Meetings in 2011," http://www.tzbsite.com/public-involvement/meetings-lists/meetings-2011.html; William Demarest, "Clarkstown Sets Two Workshops to Examine Impact of Tappan Zee Bridge Replacement Project," *New City Patch*, April 3, 2011, http://patch.com/new-york/newcity/clarkstown-sets-two-workshops-to-examine-impact-of-ta1218fac176 (workshops were originally set for January 2011.)

36 "Lt. Governor Says New TZ Bridge Is Needed," *Mid-Hudson News Network*, August 24, 2011.

37 Patrick Gallagher, "DOT Commissioner Mum on Bridge Financing," *Westchester*

Business Journal, January 13, 2012, http://westfaironline.com/18470/dot-commissioner-mum-on-bridge-financing; Khurram Saeed and Theresa Juva, "Forum Addresses Mass Transit on New Tappan Zee Bridge," *Journal News* (White Plains, NY)May 10, 2012.

38 Gallagher, "DOT Commissioner Mum on Bridge Financing."

39 Consulting team member, interview with author.

40 Ed Lucas, "The New York State Public Employees Federation Testimony of Ed Lucas Executive Board Member to the New York State Assembly Standing Committee on Transportation," December 7, 2010, http://pef.squarespace.com/storage/testimony/PEF_Testimony_DOT_2_year_Capital_Plan_Dec_2010_Testimony.pdf ; former senior aide to Governor Paterson, interview with author.

41 Gallagher, "DOT Commissioner Mum on Bridge Financing"; senior member of project team, interview with author.

42 Michael Barbaro, "Indulging an Obsession with Motors and Muscle," *New York Times*, October 28, 2010; Jonathan Mahler, "The Making of Andrew Cuomo," *New York Times*, August 11, 2010.

43 Federal Highway Administration, NYS DOT, and NYS Thruway Authority, "Tappan Zee Hudson River Crossing Project: Joint Record of Decision and State Environmental Quality Review Act Findings Statement," September 2012, http://www.newnybridge.com/documents/rod/00record-of-decision.pdf.

44 Earthtech and ARUP, *Alternatives Analysis for Hudson River Highway Crossing.*

45 FHWA official and project team member, interviews with author.

46 Project team member, interview with author.

47 "The extra-wide shoulders are not a standard highway design feature. However, they maximize the public investment by allowing space for future options on the replacement bridge," per Federal Highway Administration, NYS DOT, and NYS Thruway Authority, *Tappan Zee Hudson River Crossing Project: Final EIS*, 24–7.

48 White House Office of the Press Secretary, "Presidential Memorandum—Speeding Infrastructure Development Through More Efficient and Effective Permitting and Environmental Review," press release, August 31, 2011.

49 Dan Graves (Executive Director of the President's Council on Jobs and Competitiveness at the White House), telephone interview with author, October 28, 2011.

50 Howard Glaser e-mail to Joseph Percoco, Mika L. Rothman, Linda Lacewell, Polly Trottenberg, and John Porcari, September 13, 2011, in author's possession. Subsequent information provided by the governor's office indicated the accident rate was twice as high, not three times as high.

51 Ibid.

52 Christine Haughney, "U.S. Says It Will Expedite Approval to Replace Deteriorating Tappan Zee Bridge," *New York Times*, October 11, 2011.

53 Joanna Turner (US DOT Deputy Assistant Secretary), e-mail to Eric Beightel (US DOT Assistant Secretary for Policy), September 14, 2011, in author's possession.

54 Raymond Lahood, telephone interview with author, January 29, 2014.

55 Ibid.

56 Ibid.

57 Ibid.

58 Ibid.

59 Kate Slevin, "NYS DOT Mute on Tappan Zee Bridge Options," *Mobilizing the Region*, May 21, 2008.

60 Andrew Cuomo, "Governor Cuomo Requests Expedited Federal Approval of Tappan Zee Bridge Project," press release, October 10, 2011.

61 White House Office of the Press Secretary, "Obama Administration Announces Selection of 14 Infrastructure Projects to be Expedited Through Permitting and Environmental Review Process," press release, October 11, 2011.
62 U.S. Department of Transportation, "Rescinded Notice of Intent for the Tappan Zee Bridge/I-287 Corridor Project," October 12, 2011. (Federal transportation officials had signed off on this rescinding notice on September 26.)
63 U.S. Department of Transportation, "Notice of Intent for the Tappan Zee Hudson River Crossing Project," October 12, 2011.
64 Associated Press, "New Tappan Zee Bridge in NY Pegged at $5.2 billion," *Wall Street Journal*, October 11, 2011.
65 Federal Highway Administration, NYS DOT, and Thruway Authority, *2011 Scoping Information Packet*, October 2011, http://www.tzbsite.com/tzbsite_2/pdf-library_2/2011-10-13%20Scoping%20Information%20Packet.pdf.
66 Andrew Cuomo, interview with Fred Dicker, *Talk1300*, November 22, 2011, http://www.talk1300.com/podcast/Fred-Dicker-Live-from-the-State-Capitol/. Accessed July 24, 2012.
67 Joseph Spector, "Cuomo Makes Appeal for New Tappan Zee Bridge," *Journal News*, November 17, 2011, http://polhudson.lohudblogs.com/2011/11/17/cuomo-makes-appeal-for-new-tappan-zee-bridge (audio of speech available with article).
68 Ibid.
69 Nicole Gelinas, "Cuomo's Tappan Zee Decisions Risky Bridge-ness," *New York Post*, February 29, 2012.
70 Casey Seiler, "Cuomo: Tappan Zee a Test for Building Big," *Times Union* (Albany, NY) July 9, 2012.

Chapter 11. Public Reaction and Cuomo's Campaign (2011–2012)

1 Editorial, "New TZ More Than Just a Bridge," *Journal News* (White Plains, NY), October 11, 2011; editorial, "Keep Up Tappan Zee Pressure," *Newsday*, (Melville, NY), October 13, 2011, http://www.newsday.com/opinion/keep-up-tappan-zee-pressure-1.3243937.
2 Editorial, "Mass Transit Key in TZ project," *Journal News* (White Plains, NY), October 16, 2011; "Editorial Board Poll: Tappan Zee Bridge," *Journal News* (White Plains, NY), http://rockland.lohudblogs.com/2011/10/17/editorial-board-poll-tappan-zee-bridge; editorial, "Make Room on New Tappan Zee for Bus Rapid Transit," *Newsday* (Melville, NY), March 2, 2012.
3 Rob Astorino and Scott Vanderhoef, "Rob Astorino and Scott Vanderhoef on Tappan Zee Bridge," interview by Nick Reisman, *Capital Tonight*, December 2, 2011, http://capitaltonightny.ynn.com/2011/12/rob-astorino-and-scott-vanderhoef-on-tappan-zee-bridge (quotes were taken from video available from this website).
4 Patrick Gallagher, "Q&A with the County Executive," *Westchester Business Journal*, October 21, 2011, http://westfaironline.com/41892/qa-with-the-county-executive.
5 County of Westchester, "Astorino's Remarks at TZ Bridge Scoping Session," press release, October 26, 2011.
6 Astorino and Vanderhoef, "Rob Astorino and Scott Vanderhoef on Tappan Zee Bridge."
7 Ibid.
8 County of Westchester, "Astorino Calls on Governor to Make TZB a Priority," press release (originally issued in October 2011, the press release was subsequently revised in August 16, 2012,) http://www3.westchestergov.com/news/

all-press-releases/4054-astorino-calls-on-gov-cuomo-to-make-replacement-of-tz-bridge-a-priority; Astorino, speech to Manhattan Institute's Forum on Replacing the Tappan Zee Bridge, June 22, 2011, http://www3.westchestergov.com/images/stories/pdfs/tzbspeech.pdf.

9 Drew Fixell, telephone interview with author, January 30, 2014.

10 Ibid.

11 Mass Transit Task Force, Meeting 10 Powerpoint, October 25, 2013, http://www.newnybridge.com/documents/meetings/2013/20131025-network-analysis.pdf.

12 Fixell telephone interview.

13 Ibid.

14 Fixell, public hearing comments, Tarrytown, New York, March 1, 2012, per Federal Highway Administration, NYS DOT, and NYS Thruway Authority, *Tappan Zee Hudson River Crossing Project: Final EIS*, vol. 3.

15 Fixell telephone interview.

16 Steven P. Knowlton (Village of Nyack Trustee) and Marie Lorenzini (Village of Nyack Liaison to the Tappan Zee Bridge/287 Corridor Project), letter to Governor Cuomo, October 17, 2011, published in http://www.nyacknewsandviews.com/2011/10/vb_tzbcuomo201117.

17 Federal Highway Administration, NYS DOT, and Thruway Authority, "Tappan Zee Hudson River Crossing Project: Scoping Information Packet," October 2011, 1–2, http://www.tzbsite.com/tzbsite_2/pdf-library_2/2011-10-13%20Scoping%20Information%20Packet.pdf.

18 Noah Kazis, "Who Killed Transit on the New Tappan Zee? Feds and State DOT Won't Say," *Streetsblog*, October 24, 2011, http://www.streetsblog.org/2011/10/24/who-killed-transit-on-the-new-tappan-zee-feds-and-state-dot-wont-say.

19 New York State Department of Transportation, "Proposal for Tappan Zee Bridge & I-287 Corridor Unveiled: Team Recommends Bridge Replacement, Addition of Bus," press release, September 26, 2008; NYS DOT, NYS Thruway Authority, and Metro-North, *Scoping Summary Report*, 4–24.

20 Tom Madison, Governor Andrew Cuomo's Cabinet Meeting, February 22, 2012, PowerPoint slides and video available at http://www.newnybridge.com/documents/meetings/index.html#cabinet; Andrew Cuomo, "Statement from Governor Cuomo on the Tappan Zee Bridge," press release, August 3, 2012; NYS Thruway Authority, "Background on Previous Studies on Bus Rapid Transit (BRT)," 2012, http://www.newnybridge.com/documents/brt/index.html.

21 Kate Hinds, "Cuomo Says Mass Transit System for Tappan Zee Would Double Costs," WNYC News, July 10, 2012, http://www.wnyc.org/articles/wnyc-news/2012/jul/10/cuomo-cost-build-mass-transit; Jeffrey Zupan, "Testimony by RPA's Jeff Zupan on the Tappan Zee Bridge," February 28, 2012, http://www.rpa.org/pdf/RPA-Testimony-Tappan-Zee-2-28-12.pdf.

22 Federal Highway Administration official, interview with author.

23 Robert Conway, telephone interview with author, May 14, 2013.

24 Jeremy Smerd and Shane Dixon Kavanaugh, "Tappan into Private Investors? Gov. Andrew Cuomo Is 'Hell-Bent' on Building a New Tappan Zee Bridge," *Crain's New York*, January 22, 2012; Joseph Spector, "Cuomo's Latest Development Plan Leans on Private Investment," *Poughkeepsie Journal*, January 6, 2012.

25 Richard Kavesh, "TZB Packs SRO Crowd at Doubletree," *Nyack News and Views*, October 26, 2011, http://www.nyacknewsandviews.com/2011/10/rk_tzb_doubtree201110; Robert Knight, "Full Steam Ahead for New Tappan Zee Bridge," *Rockland County Times*, March 7, 2012.

26 Project team member, interview with author.
27 Judy Rife, "Tappan Zee Bridge Stakeholders Disinvited," *Times-Herald Record* (Middletown, NY), February 14, 2012.
28 Rockland County planner, telephone interview with author, June 6, 2012.
29 Kazis, "Who Killed Transit on the New Tappan Zee?"
30 Kate Slevin, "TZ Plan a Dud without Public Transportation," Op-Ed, *Journal News* (White Plains, NY), December 5, 2011.
31 Veronica Vanterpool, "Tappan Zee Public Hearing Testimony," March 1, 2012, http://www.tstc.org/press/2012/030112_TZ_Public_Hearing_Testimony.pdf.
32 Dani Simons, "New Radio Ads Tell Albany to Put Transit Back in the Tappan Zee," *Mobilizing the Region*, February 20, 2012, http://blog.tstc.org/2012/02/20/new-radio-ads-tell-albany-to-put-transit-back-in-the-tappan-zee.
33 Noah Kazis, "Cuomo Admin Silent as Media Questions Tappan Zee Fuzzy Math," *Streetsblog*, February 21, 2012, http://www.streetsblog.org/2012/02/21/cuomo-admin-silent-as-media-questions-tappan-zee-fuzzy-math.
34 Sam Handler, "Diverse Coalition Demands Bus Rapid Transit on New Tappan Zee," *Mobilizing the Region*, December 15, 2011, http://blog.tstc.org/2011/12/15/diverse-coalition-demands-bus-rapid-transit-on-new-tappan-zee; Khurram Saeed, "Coalition: New Tappan Zee Bridge Must Have Mass Transit," *Journal News* (White Plains, NY), December 16, 2011.
35 Kate Slevin, telephone interview with author, March 15, 2012.
36 Ross Pepe, telephone interview with author, November 30, 2012.
37 Ross Pepe, testimony at March 1, 2012, public hearing in Tarrytown, available at "Tappan Zee Hudson River Crossing Study Draft Environmental Impact Statement Public Hearing," http://www.newnybridge.com/documents/meetings/first-westchester.pdf.
38 Pepe telephone interview.
39 "Citing Public Safety, Economic Stability in New York, Coalition Rallies Supporters for New H.R. Xing in 2012," press release, ReplaceTheTZBridgeNow.org, February 28, 2012.
40 Michelle Breidenbach, "Andrew Cuomo Claims to Fight Special Interests, But Takes Millions in Donations," *Post-Standard* (Syracuse, NY), October 31, 2010.
41 Joseph Berger, "The Rise and Fall of Edward Doyle Sr.," *New York Times*, June 27, 1999.
42 Elizabeth Ganga, "Westchester BOL Campaign: Testa Endorsed by Building Trades Council," Politics on the Hudson, September 17, 2013, http://polhudson.lohudblogs.com/2013/09/17/westchester-bol-campaign-testa-endorsed-by-building-trades-council.
43 Nicholas Confessore, "Donations to Key Cuomo Ally Show a Rift among Unions," *New York Times*, June 7, 2012.
44 Al Samuels, telephone interview with author, January 31, 2014.
45 Ibid.
46 Ibid.
47 Ibid.
48 Joseph Spector, "Cuomo Makes Appeal for New Tappan Zee Bridge," *Journal News*, November 17, 2011, http://polhudson.lohudblogs.com/2011/11/17/cuomo-makes-appeal-for-new-tappan-zee-bridge (audio of speech available with article); Khurram Saeed, "Thruway Revives Claim of 50-Year Tappan Zee Bridge Life Span, But Proof Remains Elusive," *Journal News* (White Plains, NY), March 1, 2012; Madison, Governor Andrew Cuomo's Cabinet Meeting, February 22, 2012.

49 Andrew Cuomo, "President Obama Approves Governor Cuomo's Request for Expedited Federal Approval of Tappan Zee Bridge Project," press release, October 11, 2011.
50 Spector, "Cuomo Makes Appeal for New Tappan Zee Bridge."
51 Federal Highway Administration, NYS DOT, and NYS Thruway Authority, *Tappan Zee Hudson River Crossing Project: Draft Environmental Impact Statement and Section 4(f) Evaluation*, Executive Summary, January 2012, S-21.
52 Madison, Governor Andrew Cuomo's Cabinet Meeting, February 22, 2012.
53 Eliabeth Ganga, "Abinanti Weighs In On Tappan Zee; Thruway Authority Responds," *Journal News* (White Plains, NY), August 24, 2012.
54 Natalia Baage-Lord, "Proposal Would Make Tappan Zee Bridge a Walkway," *Daily Voice* (Dobbs Ferry, NY), October 19, 2011, http://rivertowns.dailyvoice.com/news/proposal-would-make-tappan-zee-bridge-walkway.
55 Peter Applebome, "What Should the Tappan Zee's Future Look Like?" *New York Times*, April 4, 2012.
56 Leading project team member, interview with author.
57 WNYC Newsroom, "Cuomo Considering Turning Tappan Zee Bridge into a Walkway," WNYC News, February 22, 2012, http://www.wnyc.org/articles/wnyc-news/2012/feb/22/cuomo-considering-turning-tappan-zee-bridge-walkway/#; Editorial, "A Tappan Zee Greenway," *New York Times*, February 29, 2012.
58 Paul Feiner, "Comments on Tappan Zee Hudson River Crossing Study Draft EIS," public hearing, Tarrytown, March 1, 2012, in Federal Highway Administration, NYS DOT, and NYS Thruway Authority, *Tappan Zee Hudson River Crossing Project: Final EIS,* vol. 3.
59 Milagros Lecuona, "Comments on Tappan Zee Hudson River Crossing Study Draft EIS," March 1, 2012, in Federal Highway Administration, NYS DOT, and NYS Thruway Authority, *Tappan Zee Hudson River Crossing Project: Final EIS*, vol. 3.
60 Dave Zornow, "Tappan Zee Bridge Greenway: Dead Before Arrival," *Nyack News and Views*, March 18, 2012.
61 Nick Reisman, "How to Pay for the Tappan Zee Bridge?" *Capital Tonight* on YNN, October 12, 2011, http://www.capitaltonight.com/2011/10/how-to-pay-for-the-tappan-zee-bridge (video available on website). Accessed July 24, 2012.
62 Associated Press, "New Tappan Zee Bridge in NY Pegged at $5.2 billion," *Wall Street Journal*, October 11, 2011.
63 Jon Campbell, "Cuomo Looks to Pensions to Help Pay for New Bridge," *Poughkeepsie Journal*, November 22, 2011.
64 Federal Highway Administration, "TIFIA Defined," https://www.fhwa.dot.gov/ipd/tifia/defined.
65 Madison, Governor Andrew Cuomo's Cabinet Meeting, February 22, 2012; Judy Rife, "Toll Hike Likely on New Tappan Zee Bridge: Rates Will Be Linked to Cost of Project," *Times Herald-Record* (Middletown, NY), March 6, 2012.
66 Jim O'Grady, "NY Gov Cuomo Offers No Details on Tappan Zee Bridge Funds," *Transportation Nation*, WNYC, May 3, 2012, http://www.wnyc.org/blogs/transportation-nation/2012/may/03/ny-gov-cuomo-sounds-can-do-about-tappan-zee-bridge-but-funding-remains-unclear.
67 Kate Hinds, "Cuomo Says Mass Transit System for Tappan Zee Would Double Costs," WNYC News, July 10, 2012, http://www.wnyc.org/story/221810-cuomo-cost-build-mass-transit/; Kate Hinds, "Financial Plan for Tappan Zee Bridge Probably Won't Come Until August," *Transportation Nation*, WNYC, May 10, 2012, http://www.wnyc.org/story/284568-financial-plan-for-tappan-zee-bridge-probably-wont-come-until-august/.

68 R. Knight, "Full Steam Ahead for New Tappan Zee Bridge," *Rockland County Times*, March 7, 2012.
69 Rockland County official, conversation with author, South Nyack, February 28, 2012.
70 Scott Vanderhoef, interview with author, Pomona, May 15, 2012.
71 Elizabeth Ganga, Khurram Saeed, and Theresa Juva, "Tappan Zee Bridge: Feds Say Come Back Later on State's $2B Loan to Fund Replacement," *Journal News* (White Plains, NY), April 26, 2012; Theresa Juva-Brown and Khurram Saeed, "County Executives Want Info, Leave Tappan Zee Bridge Fate Unclear," *Journal News* (White Plains, NY), July 8, 2012.
72 Vanderhoef interview.
73 John Wagner, "Debate Over Tappan Zee Bridge Replacement Continues," YNN, July 10, 2012, http://hudsonvalley.ynn.com/content/top_stories/591110/debate-over-tappan-zee-bridge-replacement-continues.
74 Ed Buroughs, letter to Michael Anderson, March 30, 2012; Scott Vanderhoef, letter to Michael Anderson, March 30, 2012.
75 Vanderhoef interview.
76 Jon Campbell, "Astorino, Cuomo Aide Spar Over Replacement for Hold Your Breath Bridge," Politics on the Hudson, *Journal News* (White Plains, NY), July 10, 2012.
77 Governor's June 2012 Schedule and July 2012 Schedule, available at http://programs.governor.ny.gov/citizenconnects/content/past-schedule.
78 Marathon Strategies, "Strategic Communications," http://www.marathonstrategies.com/what; Politico, "Arena Profile: Phil Singer," http://www.politico.com/arena/bio/phil_singer.html.
79 Kevin Zawacki, "New Tappan Zee Bridge to Include Rush-Hour Bus Lanes," *Nyack-Piermont Patch*, June 29, 2012, http://nyack.patch.com/articles/new-tappan-zee-bridge-to-include-rush-hour-bus-lanes.
80 Project team member, interview with author.
81 Federal Highway Administration, NYS DOT, and NYS Thruway Authority, *Tappan Zee Hudson River Crossing Project: Final EIS*, 4–19.
82 Governor's July 2012 Schedule, available at http://programs.governor.ny.gov/citizenconnects/content/past-schedule.
83 Fredric U. Dicker, "Cuomo's Top Aide to Manage $5B Tappan Zee Bridge Project," *New York Post*, July 9, 2012.
84 Senior state official, interview with author.
85 Project team member, interview with author.
86 NYS Thruway Authority, "Public Meetings—Presentations and Minutes (2000–Present)," http://www.newnybridge.com/documents/meetings/index.html.
87 Janie Rosman, "Area Residents Pepper Officials with Questions about Tappan Zee," *Hudson Independent*, November 2, 2012.
88 NYS Thruway Authority, "2012 Press Releases," http://www.thruway.ny.gov/news/pressrel/2012/index.html. Accessed November 16, 2013.
89 John Dyer, "Cuomo Sees Tappan Zee As a Model for Better Government," *Newsday* (Melville, NY), October 17, 2012, http://newyork.newsday.com/westchester/cuomo-sees-tappan-zee-as-a-model-for-better-government-1.4124171. Accessed March 16, 2013.
90 Federal Highway Administration official, interview with author.
91 Theresa Juva-Brown and Khurram Saeed, "New Tappan Zee Bridge or Not, Tolls Will Rise," *Journal News*, August 4, 2012; NYS Thruway Authority, "Toll Options," August 2012, http://www.newnybridge.com/documents/study-documents/toll-options.pdf.
92 Vanderhoef interview.

93 Federal Highway Administration, NYS DOT, and NYS Thruway Authority, "Tappan Zee Hudson River Crossing Project: Scoping Information Packet."

94 A project team member provided the author with an estimate of the number of pages; no one actually counted the number of pages; Freeman Klopott, "Cuomo Tappan Zee Plan Gets U.S. Approval in Obama Fast Track," *Bloomberg News*, September 25, 2012, http://www.bloomberg.com/news/2012-09-25/cuomo-s-tappan-zee-plan-gets-u-s-approval-in-obama-fast-track.html.

95 U.S. Government, "Federal Infrastructure Projects Permitting Dashboard," http://www.permits.performance.gov/projects/tappan-zee-bridge-replacement.

96 Federal official, interview with author.

97 Project team member, interview with author.

98 AKRF consultant, interview with author.

99 Conway telephone interview.

100 Project team member, interview with author.

101 Project team consultant, interview with author.

102 Conway telephone interview.

103 AECOM consultant, interview with author.

104 William Janeway (NYS Department of Environmental Conservation), letter to Michael Anderson, April 17, 2008 (in NYS DOT, NYS Thruway Authority, and Metro-North, *Scoping Comments Report*, appendix B.)

105 Federal Highway Administration official, interview with author.

106 Karen Rae (presentation, panel on Unique Challenges of Mega-Project Delivery, Transportation Research Board Annual Meeting, Washington, DC, January 13, 2014).

107 Henry L. Goldberg, "Infrastructure Investment Act Implements Design Build in New York," February 2012, http://www.goldbergconnolly.com/pages/publications/articles/february_legal_log2012.pdf; New York State Legislature, Senate Bill S50002–2011 and Assembly Bill A40002–2011.

108 Debra Rubin and Aileen Cho, "Procuring N.Y.'s $5-Billion Tappan Zee Bridge Will Be Tight," *ENR: Engineering News-Record*, February 20, 2012.

109 Thomas Zambito and John Dyer, "Low Tappan Zee Bid Could Knock $5 Off Toll," *Newsday*, December 6, 2012, http://newyork.newsday.com/westchester/low-tappan-zee-bid-could-knock-5-off-toll-1.4295314; Roberto Cruz, "Tappan Zee Design Presented to Thruway Authority," *Legislative Gazette*, December 10, 2012.

110 Judy Rife, "New Tappan Zee Announced," *Times Herald-Record*, December 18, 2012.

111 Fredric U. Dicker, "Trump Wants to Fix Tappan Zee for Peanuts," *New York Post*, December 23, 2013.

Chapter 12. Lost Opportunities and Wasted Resources

1 Federal Transit Administration and New Jersey Transit, *Access to the Region's Core: Final Environmental Impact Statement [EIS].*

2 Sam Roberts, "More People Commute to Manhattan Than Any Other County," *New York Times*, March 5, 2013; David Lombino, "Senators Call on Government to Help Fund City-Jersey Rail Link," *New York Sun*, May 16, 2006.

3 Tricia Tirella, "Groundbreaking Begins on $8.7B Transit Project through Palisades," *Hudson Reporter*, June 14, 2009.

4 Federal Transit Administration and New Jersey Transit, *Access to the Region's Core: Final EIS.*

5 Federal Transit Administration, *Annual Report on Funding Recommendations: Fiscal*

Year 2010, A-121; NYS DOT, NYS Thruway Authority, and Metro-North, *Transit Mode Selection Report*, table 5–3.

6 Federal Transit Administration, *Annual Report on Funding Recommendations: Fiscal Year 2010*, A-121; NYS DOT, NYS Thruway Authority, and Metro-North, *Scoping Summary Report*.

7 Peter Derrick, interview with author, Riverdale, New York, May 26, 2011.

8 Jeff Zupan, memo to Janet Mainiero (Metro-North) and Chris Waite (New York State Thruway Authority), April 28, 2004, in author's possession.

9 Janet Mainiero, telephone interview with author, January 28, 2013.

10 Based on author's interviews with five different transportation officials.

11 Tom Schulze, interview with author, Newark, New Jersey, May 24, 2011.

12 Ernie Salerno, telephone interview with author, April 14, 2012.

13 Chris Boylan, e-mail to Michael Mannix, July 19, 2006, in author's possession.

14 Federal Transit Administration and New Jersey Transit, *Access to the Region's Core: Final EIS*, ES-6.

15 *Star-Ledger* staff, "Hudson River Tunnel Project Is Officially Canceled by Gov. Christie," *Star-Ledger*, October 7, 2010.

16 Peggy Ackermann, "Gov. Christie Criticizes New York for Not Contributing to Halted ARC Tunnel Project," *Star-Ledger*, October 28, 2010. According to author's interviews with Governor David Paterson and MTA executive director Lee Sander, Governor Christie never asked New York's governor or the MTA to pay for ARC.

17 MTA official, interview with author.

18 Martin Robins, interview with author, Westfield, New Jersey, January 7, 2012.

19 Former transportation official, interview with author.

20 MTA official, interview with author.

21 Richard Perez-Pena, "Plan Edges L.I.R.R. Closer To a Grand Central Link," *New York Times*, April 25, 1996; George E. Pataki, "Master Links: A Regional Transportation Vision for the 21st Century," undated, in author's possession (this report was prepared by the MTA but never released to the public).

22 Alfonso A. Castillo, "East Side Access Completion Date Extended—Again," *Newsday*, January 27, 2014.

23 Thomas DiNapoli, "MTA's East Side Access Project 10 Years Late and $4.4 Billion Over Budget," press release, March 6, 2013.

24 MTA, New Jersey Transit, and Port Authority, "Access to the Region's Core Summary Report," 2003, 34.

25 Don Eisele, telephone interview with author, August 27, 2011,

26 Federal Transit Administration and New Jersey Transit, *Access to the Region's Core: Final EIS*, ES-5; MTA, New Jersey Transit, Port Authority, "Access to the Region's Core (Phases 1 and 2): Milestone Summary Report."

27 George Haikalas, interview with author, New York, March 11, 2011.

28 George Pataki, interview with author, Rockefeller Center, December 14, 2011.

29 Virgil Conway, interview with author, New York, April 5, 2011; Lou Tomson, telephone interview with author, August 5, 2011.

30 Federal Highway Administration and NYS DOT, *I-287/Cross Westchester Expressway: Final Design Report/Final EIS*, VI-449.

31 James Yarmus, interview with author, New City, New York, May 17, 2011.

32 Federal Highway Administration official, interview with author.

33 Joseph C. Ingraham, "Thruway Bridge Will Open Today; Full Toll Length of Road to Be Put into Operation," *New York Times*, December 15, 1955.

34 Chris Waite, e-mail to author, October 27, 2012.

35 NYS Thruway Authority, "State Holds Public Hearings On Building a New Tappan Zee Bridge As Part of Accelerated Construction Schedule," press release, March 1, 2012.

36 Federal Highway Administration, NYS DOT, and NYS Thruway Authority, *Tappan Zee Hudson River Crossing Project: Final EIS*, A-6; Judy Rife, "New Deck for the Tappan Zee: Completion of Replacement to Take 18 Months," *Times Herald-Record* (Middletown, NY), May 17, 2011.

Conclusion

1 Marty Huss, telephone interview with author, March 7, 2011.

2 Flyvbjerg, *Policy and Planning for Large Infrastructure Projects*, 9.

3 Wachs, "Ethics and Advocacy."

4 Lou Tomson, telephone interview with author, August 5, 2011.

5 John Shafer, telephone interview with author, September 8, 2011.

6 Maryanne Gridley, telephone interview with author, August 5, 2011.

7 Carrie Laney, telephone interview with author, November 28, 2011.

8 Altshuler and Luberoff in *Mega-Projects* explain the importance of public sector champions.

9 Panetta, *The Tappan Zee Bridge*, 54; Jeremy Smerd and Shane Dixon Kavanaugh, "Tappan into Private Investors? Gov. Andrew Cuomo Is 'Hell-Bent' on Building a New Tappan Zee Bridge," *Crain's New York*, January 22, 2012.

Bibliography

Citations refer to materials located in four different archives: New York Metropolitan Transportation Council, New York State Department of Transportation, New Jersey Transit, and Westchester County Department of Public Works and Transportation. The NYS DOT materials were obtained from the Region 8 office in Poughkeepsie, New York (hereafter NYS DOT Archives). The New York Metropolitan Transportation Council materials are available to the public at the council's Manhattan library (hereafter NYMTC Archives). New Jersey Transit's Access to the Region's Core materials are stored at its headquarters in Newark (hereafter NJ Transit Archives). Westchester County materials are stored in the Department of Public Works and Transportation's offices in Mount Vernon, New York (hereafter County Archives).

Adler, Thomas, Willard Ristau, and Stacey Falzarano. "Traveler Reactions to Congestion Pricing Concepts for New York's Tappan Zee Bridge." *Transportation Research Record: Journal of the Transportation Research Board* 1659, no. 1 (1999): 87–96.

Altshuler, Alan, and David Luberoff. *Mega-Projects: The Changing Politics of Urban Public Investment.* Washington, DC: Brookings Institution, 2003.

American Association of State Highway and Transportation Officials [AASHTO]. *Transportation—Invest in our Future: Accelerating Project Delivery.* Washington, DC: 2007.

Beauregard, Robert A. *When America Became Suburban.* Minneapolis: University of Minnesota Press, 2006.

Benjamin-Bothwell, Sharon. "An Intervention Addressing a Wicked Problem: Leadership Tension between Executives and Boards in National Nonprofit Organizations." PhD diss., Union Institute and University Graduate College, 2003.

Caro, Robert A. *The Power Broker: Robert Moses and the Fall of New York.* New York: Knopf, 1974.

Conklin, Jeff. *Dialogue Mapping: Building Shared Understanding of Wicked Problems.* Chichester, Eng.: John Wiley & Sons, 2005.

Cuomo, Andrew. *All Things Possible: Setbacks and Success in Politics and Life.* New York: HarperCollins, 2014.

Danielson, Michael N., and Jameson W. Doig. *New York: The Politics of Urban Regional Development.* Berkeley: Institute of Governmental Studies by University of California Press, 1982.

Doig, Jameson W. "If I See a Murderous Fellow Sharpening a Knife Cleverly . . . the Wilsonian Dichotomy and the Public Authority Tradition." *Public Administration Review* (1983): 292–304.

Dollery, Brian E., and Joe L. Wallis. *The Political Economy of Local Government: Leadership, Reform, and Market Failure*. Cheltenham, UK: Edward Elgar, 2001.

Downs, Anthony. *Stuck in Traffic: Coping with Peak-Hour Traffic Congestion*. Washington, DC: Brookings Institution, 1992.

———. *Still Stuck in Traffic: Coping with Peak-Hour Traffic Congestion*. Washington, DC: Brookings Institution Press, 2004.

Earthtech and ARUP. *Alternatives Analysis for Hudson River Highway Crossing* (prepared for NYS DOT, Thruway Authority, Metro-North). July 2007. http://www.tzbsite.com/tzb-library/pdf-library/pdf-HR-reports/20070702-Hudson%20River-Highway-Crossing.pdf.

Federal Highway Administration and NYS DOT. *I-287/Cross Westchester Expressway: Final Design Report/Final Environmental Impact Statement*. June 1997. County Archives.

Federal Highway Administration, NYS DOT, and NYS Thruway Authority. *Tappan Zee Hudson River Crossing Project: Final Environmental Impact Statement and Section 4(f) Evaluation*. July 2012. http://www.newnybridge.com/documents/feis.

Federal Transit Administration. *Annual Report on Funding Recommendations: Fiscal Year 2010 New Starts, Small Starts, and Paul S. Sarbanes Transit in Parks Program*. 2009. http://www.fta.dot.gov/12304_9672.html.

———. *Annual Report on New Starts: Proposed Allocation of Funds for Fiscal Year 2007, Report of the Secretary of Transportation to the United States Congress Pursuant to 49 U.S.C. 5309(k)(1)*, 2006. http://www.fta.dot.gov/12304_2639.html.

Federal Transit Administration and New Jersey Transit. *Access to the Region's Core: Draft Environmental Impact Statement in Hudson County (NJ) and New York (NY)*. January 2007. NJ Transit Archives.

———. *Access to the Region's Core: Final Environmental Impact Statement in Hudson County (NJ) and New York (NY)*. October 2008. NJ Transit Archives.

Fein, Michael R. *Paving the Way: New York Road Building and the American State, 1880–1956*. Lawrence: University Press of Kansas, 2008.

Ferrandino & Associates. *Interstate 287 Monorail Feasibility Study: Prepared for the Construction Industry Foundation*. Mount Vernon, NY. October 1986.

Flyvbjerg, Bent. *Policy and Planning for Large Infrastructure Projects: Problems, Causes, Cures*. Policy Research Working Paper 3781. World Bank, 2005.

———. "Truth and Lies about Megaprojects." Inaugural speech for professorship and chair at Faculty of Technology, Policy, and Management, Delft University of Technology, 2007.

Frug, G. E. "Beyond Regional Government." *Harvard Law Review* 115 (2002): 1763–1836.

Garreau, Joel. *Edge City: Life on the New Frontier*. New York: Anchor Books, 1992.

Glaeser, Edward L. *Urban Colossus: Why Is New York America's Largest City?* Working Paper 11398. National Bureau of Economic Research, 2005. http://www.nber.org/papers/w11398.

Goetz, Andrew R., and Joseph S. Szyliowicz. "Revisiting Transportation Planning and Decision Making Theory: The Case of Denver International Airport." *Transportation Research Part A: Policy and Practice* 31, no. 4 (1997): 263–280.

Holguín-Veras, José, and Robert E. Paaswell. "New York Regional Intermodal Freight Transportation Planning: Institutional Challenges." *Transportation Law Journal* 27 (2000): 453–473.

Jackson, Kenneth T. *Crabgrass Frontier: The Suburbanization of the United States*. New York: Oxford University Press, 1985.

———. "Foreword." In *Westchester: The American Suburb*, edited by Roger G. Panetta, vii–ix. New York: Fordham University Press, 2006.

Kantor, Paul. "The Coherence of Disorder: A Realist Approach to the Politics of City Regions." *Polity* 42, no. 4 (2010): 434–460.
Ketchum, E. P. (District Engineer). *Report on Application for Approval of Plans of Bridge Proposed to be Constructed Across Hudson River, between Nyack, Rockland County, and Tarrytown, Westchester County, N.Y.* Army Corps of Engineers, March 29, 1951. Office of the U.S. Coast Guard First District, New York.
Levy, Sidney M. *Build, Operate, Transfer: Paving the Way for Tomorrow's Infrastructure.* New York: John Wiley & Sons, 1996.
Mallett, William, and Linda Luther. *Accelerating Highway and Transit Project Delivery: Issues and Options for Congress.* Washington, DC: Congressional Research Service, 2011.
Marlin, George J. *Squandered Opportunities: New York's Pataki Years.* South Bend: IN: St. Augustine's Press, 2006.
Metropolitan Transportation Authority, New Jersey Transit, and Port Authority of New York and New Jersey. "Access to the Region's Core Summary Report." 2003. NJ Transit Archives.
———. "Access to the Region's Core (Phases 1 and 2): Milestone Summary Report." May 1999. NJ Transit Archives.
Milikowsky, Brina. *Falling Apart and Falling Behind.* Building America's Future Educational Fund, 2011 http://www.bafuture.org/pdf/Building-Americas-Future-2012-Report.pdf.
Muzzio, Douglas. "Politics and the News Media in the Empire State." In *The Oxford Handbook of New York State Government and Politics*, edited by Gerald Benjamin. New York: Oxford University Press, 2012.
New York Metropolitan Transportation Council [NYMTC]. "Tappan Zee Corridor Study: Phase 1 Report." November 1, 1984. NYMTC Archives.
———. "Tappan Zee Corridor Study: Phase III Report." 1985. NYMTC Archives.
NYS Commission on State Asset Maximization. *Final Report.* June 2009. http://www.bcnys.org/inside/transport/2011/StateAssetMaximaztion-final-report2009.pdf.
NYS DOT. "I-287/Cross Westchester Expressway Development Study: Final Project Development Report." April 1988. NYS DOT Archives.
———. "Preliminary Problem Definition and Project Proposal: Interstate 4R Rehabilitation, Restoration, Resurfacing, and Safety-Related Improvements to the Cross Westchester Expressway (I-287)." April 1983. NYS DOT Archives.
———. "Tappan Zee Corridor Study: Final Recommendations." May 1987. NYMTC Archives.
NYS DOT, NYS Thruway Authority, and Metro-North. *Alternatives Analysis for Rehabilitation and Replacement of the Tappan Zee Bridge.* March 2009. http://www.newnybridge.com/documents/study-documents/old-study-docs/alt-analysis/
———. *Alternatives Analysis Report: TZB / I-287 Environmental Review.* January 2006. http://www.tzbsite.com/tzb-library/study-documents/other-documents/alternatives-analysis-200601.html
———. *Scoping Comments Report (Tappan Zee Bridge/I-287 Corridor Project Environmental Impact Statement).* May 2009. http://www.tzbsite.com/tzb-library/study-documents/scoping-closure/scoping-comments-report_2009.html
———. *Scoping Summary Report (Tappan Zee Bridge/I-287 Corridor Project Environmental Impact Statement).* May 2009. http://www.tzbsite.com/tzb-library/study-documents/scoping-closure/scoping-summary-report_2009.html
———. *Transit Mode Selection Report.* May 2009. http://www.tzbsite.com/tzb-library/study-documents/level-3/L3-transit-mode-selection-report2009.html
Paaswell, Robert E., and Joseph Berechman. "Models and Realities: Choosing Transit Projects for New York City." In *Policy Analysis of Transport Networks: Transport and Mobility Series*, edited by M. S. van Geenhuizen, Aura Reggiani, and Piet Rietveld, 77–100. Aldershot, UK, and Burlington, VT: Ashgate, 2007.

Panetta, Roger G. *The Tappan Zee Bridge and the Forging of the Rockland Suburb.* New City, NY: Historical Society of Rockland County, 2010.

Parsons Brinckerhoff. *Feasibility and Benefit-Cost Study of Trans-Hudson, Cross-Westchester and Stewart Airport Rail Links* (prepared for Metro-North). February 1994. Metro-North offices, New York.

Pataki, George E., and Daniel Paisner. *Pataki: An Autobiography.* New York: Viking, 1998.

Petrocelli, Joseph, Richard Peters, Wayne Ugolik, Sy Schulman, and Robert Ancar. *I-287 Suffern-Port Chester Corridor HOV/TSM Action Plan: 1988–1989* (prepared for the Suffern-to-Port Chester Corridor HOV/TSM Task Force). September 1989. NYS DOT Archives.

Pickrell, D. H. "A Desire Named Streetcar: Fantasy and Fact in Rail Transit Planning." *Journal of the American Planning Association* 58, no. 2 (1992): 158–176.

Plotch, Philip Mark. "What's Taking So Long? Identifying the Underlying Causes of Delays in Planning Transportation Megaprojects in the United States." *Journal of Planning Literature* (January 2015).

Ravitch, Richard. *Report of the Lieutenant Governor on New York State's Transportation Infrastructure.* November 2010. http://www.rockinst.org/pdf/budgetary_balance_ny/2010-11-17-LG-transportation.pdf.

Ritchey, Tom. "Wicked Problems. Structuring Social Messes with Morphological Analysis." Swedish Morphological Society, 2005. http://www.swemorph.com/wp.html.

Rittel, H. W. J., and M. M. Webber. "Dilemmas in a General Theory of Planning." *Policy Sciences* 4, no. 2 (1973): 155–169.

Rockland County Planning Board. *Rockland County: River to Ridge—A Plan for the 21st Century.* 2001.

Sierra Club. "The Dark Side of the American Dream: The Costs and Consequences of Suburban Sprawl." Sierra Club. 1998. http://vault.sierraclub.org/sprawl/report98.

Smith, Robert Gillen. *Ad Hoc Governments: Special Purpose Transportation Authorities in Britain and the United States.* Beverly Hills, CA: Sage Publications, 1974.

Tri-State Transportation Campaign. *Citizens Action Plan: A 21st Century Transportation System.* New York, 1994.

Vollmer Associates, Linda Spock, Edwards & Kelcey, Allee King Rosen & Fleming, and Zetlin Strategic Communications. "Final Report for Long Term Needs Assessment." (Prepared for Governor's I-287 Task Force). April 2000. http://www.tzbsite.com/tzb-library/pdf-library/pdf-other-documents/long-term-needs-alts-analysis-200004.pdf.

Wachs, Martin. "Ethics and Advocacy in Forecasting for Public Policy." *Business & Professional Ethics Journal* 9, no. 1 (Spring 1990): 141.

Winders, Edward E. "Public Authorities in New York State: The Interdependent Governmental Roles of the New York State Thruway Authority." PhD diss., University at Albany, State University of New York, 1998.

Wolf, Donald E. *Crossing the Hudson: Historic Bridges and Tunnels of the River.* New Brunswick, NJ: Rutgers University Press, 2010.

Index

Page numbers in *italics* indicate illustrations. Numbers followed by n refer to endnotes.

About the Author

PHILIP MARK PLOTCH, PhD, is an assistant professor and director of the master's program in public administration at Saint Peter's University in Jersey City, New Jersey. He has also taught transportation planning at Hunter College's Department of Urban Affairs and Planning in New York.

Dr. Plotch has played a leading role in improving the New York metropolitan area's infrastructure. As the director of World Trade Center Redevelopment and Special Projects at the Lower Manhattan Development Corporation, he helped lead the nation's effort to rebuild Lower Manhattan after the attacks of September 11, 2001. In his previous positions as manager of planning and manager of policy at the headquarters of New York's Metropolitan Transportation Authority, he planned multibillion-dollar projects, developed emergency response procedures, and created strategic business plans.

Dr. Plotch received his bachelor's degree from the State University of New York at Albany, master's degree in urban planning from Hunter College, and PhD in public and urban policy from the Milano School of International Affairs, Management, and Urban Policy at the New School for Public Engagement. He has enjoyed living and working on both sides of the Hudson River.